Cave and Karst Systems of the World

Series editor
James W. LaMoreaux, Tuscaloosa, USA

For further volumes:
http://www.springer.com/series/11987

Max P. Cooper • John E. Mylroie

Glaciation and Speleogenesis

Interpretations from the Northeastern United States

Springer

Max P. Cooper
Department of Geosciences
Mississippi State University
Starkville, MS
USA

John E. Mylroie
Department of Geosciences
Mississippi State University
Starkville, MS
USA

Cave and Karst Systems of the World
ISBN 978-3-319-36540-4
ISBN 978-3-319-16534-9 (eBook)
DOI 10.1007/978-3-319-16534-9

Springer Cham Heidelberg New York Dordrecht London

Softcover reprint of the hardcover 1st edition 2015

Printed on acid-free paper

Springer International Publishing AG Switzerland is part of Springer Science+Business Media (www.springer.com)

Foreword

John Mylroie's 1977 dissertation on caves and karst of the Helderberg Plateau in Schoharie County, New York has long been a bible for cave investigators in that area. Now, he and his student Max Cooper—who has investigated maze caves in the northeastern United States—have produced a book which incorporates the many findings from ensuing decades, and which emphasizes the profound effect of glaciation on cave and karst development. Not limited to Schoharie County, this book covers New York and New England—the entire Northeast.

Cave studies began in the Northeast in the 1950s, with systematic tabulation of cave and karst features, and initial mapping. This has continued to the present, as new caves are discovered and mapped. But in the 1960s, scientific studies of cave development (speleogenesis) began. Insights from this have resulted in the discovery of major caves like Barrack Zourie and Carthusian caves, Thunder Hole, and Vermonster. Informal digging groups have followed the water (and speleogenetic principles) to open many new caves.

To some geologists, glaciation is an annoyance, since it blankets large areas of bedrock and has eroded strata away. But to other geologists, glaciation is "the frosting on the cake" and a fascinating field of study. Cavers and speleologists might have similar feelings. Some shallow caves were destroyed by glaciers, and glacial sediment now blocks many cave passages. But glaciation also caused many new caves to form, and within preglacial caves it modified old passages and created new ones.

As just one example of glacial effects, consider drumlins—cigar-shaped hills composed of sediments of all sizes bulldozed under an advancing glacier and aligned with ice flow. A drumlin blocked the preglacial cave stream exit of Schoharie Caverns (Schoharie County, NY) and caused the stream to form a small tapoff exit passage and, during flood conditions, to force an exit up through the glacial till. Artificial excavation in 1958 resulted in the present spacious entrance, although slumping is starting to re-bury it. Multiple streams running off a drumlin east of Howes Cave, also in Schoharie County, have formed small pit caves in the underlying limestone. As the drumlin erodes back, the cave entrances have also migrated back following the edge of the drumlin. And finally, in eastern Schoharie County two adjacent drumlins have a narrow area of exposed limestone between them. New caves are forming along joints in this valley.

There are many other ways glaciation has had major effects on cave development, and *Glaciation and Speleogenesis: Interpretations from the Northeastern United States* discusses them in detail. The significant effects of glaciation on the development of non-solutional caves such as talus, fracture, and sea caves are covered as well. Each of the several karst regions of the Northeast is treated separately, with selected caves used as case studies.

Glaciation and Speleogenesis: Interpretations from the Northeastern United States also includes a geologic history of the Northeast, a summary of the cave-forming rock units in each region, and the prospects for learning paleoclimatic data from cave deposits.

Glaciation and Speleogenesis: Interpretations from the Northeastern United States is thorough, well-illustrated, and readable, and should be in the library of every serious northeastern caver as well as karst scientists worldwide who are interested in the effects of glaciation on cave development.

January 2015 Chuck Porter

Preface

Thoughts on the effects of glaciation and karst have evolved from the late 1960s to today. The effects of continental glaciation on karst have been primarily studied in Britain (e.g., Waltham 1974), Canada (e.g., Ford 1983, 1987), Norway (e.g., Lauritzen 1981), and the northeastern United States, particularly in the Helderberg Plateau of New York (e.g., Mylroie 1977). Alpine glaciation has additionally been studied for its effects on karst in the Alps, the Pyrenees, and the Canadian Rockies (Ford and Williams 2007). The studies on continental and alpine glaciation show varying effects between the two styles of glaciation, with alpine karst development having high hydraulic gradients and strong structural control that may be missing in lowland carbonate plateaus. Early thoughts on karst and glaciation were that glaciers removed the vast majority of pre-glacial caves and karst, and those larger cave systems, which could not be established in the time since deglaciation, were anomalous (e.g., Jennings 1971). In the early 1970s in Britain it was demonstrated that there are numerous cave systems that survived glaciation (Waltham 1974), albeit these caves are commonly filled with glacial sediments. Work in the Canadian Rockies also demonstrated the survival of pre-glacial cave systems in alpine glaciated settings (Ford 1983). The overall problem is best stated by Lauritzen and Skoglund (2013, p. 365): "Few landscapes reflect a single process; polygenetic development is the rule rather than the exception. In landscapes with a very strong glacial overprint, it is difficult to know by inspection whether karstification proceeded alongside glaciation, in intervals between glaciations, or in both situations."

Following the work in Britain much research was done in the 1970s in Schoharie and Albany counties in New York State of the northeastern United States looking for, and studying large, pre-glacial cave systems (Baker 1973, 1976; Kastning 1975; Palmer 1976; Mylroie 1977, 1984). From these studies efforts shifted toward locating large, pre-glacial cave systems, with less attention paid to the small, presumed post-glacial caves. These efforts located several large, pre-glacial cave systems including those of scientific interest such as Barrack Zourie Cave (Dumont 1995).

Recent and current work in Norway (e.g., Faulkner 2008; Lauritzen and Skoglund 2013) has demonstrated that caves can form in the time since deglaciation, supporting the idea that caves can form to traversable size in 10,000 years (Palmer 1991), and that deglaciation may have stimulating effects on the formation of caves. Following this idea work continued in New York to demonstrate post-glacial origins for various caves (Cooper and Mylroie 2014). These studies have reintroduced the idea of post-glacial caves, and show that the current Holocene landscape of the northeastern United States includes a mixture of pre-glacial and post-glacial caves.

Several recent books have been published detailing karst regions that have undergone alpine glaciation (e.g., Wildberger and Preiswork 1997) and continental glaciation (e.g., Waltham and Lowe 2013; Lauritzen 2010). The recent book *Caves and Karst of the Yorkshire Dales* (Waltham and Lowe 2013) details one of the world's classic glaciated karst regions, and includes a chapter on glaciation. For New York and New England, however, little has been done chronicling the impact this region has had on the evolving thoughts on glaciated karst. The two sources mentioning this region are a review paper by Mylroie and Mylroie (2004), and two chapters in *Caves and Karst of the USA* (Palmer and Palmer 2009) on New York (Engel 2009) and New England (Porter 2009). Other books on parts of the region, including *Vermont Caves: A geologic and historical guide* (Quick 2010), do not explicitly detail the studies that evolved thoughts on glaciated karst. The currently published works also do not include the recent studies that have taken place such as Faulkner (2009), Perzan et al. (2014), Weremeichik and Mylroie (2014), and Cooper and Mylroie (2014).

New York and New England therefore present an opportunity to discuss the caves of the region, as well as the evolving thoughts on continentally glaciated karst and how this region has influenced these thoughts. Within the boundaries of a 100 km-radius circle centered on Albany, New York, exist glaciated caves and karst formed in Precambrian marbles, Cambro-Ordovician marbles, flat-lying Cambrian through Devonian limestones, and Cambrian through Devonian limestones that are highly deformed. The karst exists from sea level through 1 km elevation, from areas once covered by traditional continental ice sheets in the Mohawk-Hudson lowlands to late-stage alpine glaciation in the Adirondack Mountains. No other place on earth displays the variety of carbonate rock types, tectonics, and glaciation as seen in the northeastern United States.

The Northeastern region offers chronicles of scientific works, as well as amateur works such as those published in the caving journal *The Northeastern Caver*. The region has a rich caving community that actively looks for new caves, and has recently discovered large systems such as Barrack Zourie and Thunder Hole in Schoharie County, New York.

This book will explore the scientific and amateur works on karst and caves in the northeastern United States, paying attention to the interaction of glaciation on the karst, the geologic controls on karst of the region, the human interaction with the caves by cavers and scientists, and the evolution of thought processes regarding glaciation and karst. For a detailed look at the physical and chemical processes in action during the glaciation of karst, the reader is referred to the exceptional treatise on the subject by Lauritzen and Skoglund (2013).

To explore these facets this book presents a brief review of the effects of continental and alpine glaciation including their effects on karst, the geology of the northeastern United States and its evolution through time, with details pertinent to karst landscape evolution, and a localized view of the northeastern US cave and karst studies comparing the karst regions of New York and New England to other worldwide examples such as Britain and Norway. Further, the book explores case studies of cave systems in the variety of settings within the region including karst developed in:

(1) the Siluro-Devonian limestones of central New York including the gently tilted Helderberg Plateau and the deformed north-south band of Helderberg Group outcrops adjacent to the Hudson River,
(2) the Cambro-Ordovician limestones of the northern New York and western Vermont lowlands,
(3) the Precambrian, Cambrian, and Ordovician marbles of New England and eastern New York,
(4) the Precambrian Grenville Marble of the Adirondacks.

Additionally, various non-dissolution caves are discussed in a chapter on pseudokarst, relating pseudokarst development to glaciation. These case studies each include various scientific and amateur studies performed in the region, and the human interactions with caves. Each case study also covers the landscape evolution of the settings from the pre-glacial caves to post-glacial.

Several conventions will be used in this book pertaining to terminology and the use of units. Units will be reported as given in their original database, with metric conversions following, if applicable. Karst terminology will on occasion reflect that commonly used in Britain, as many terms relating to glaciated karst originated there. An example of this includes "grike and glint" as opposed to "cutter and pinnacle" for dissolutionally opened joints and the residual blocks left between. Additional terminology preferences include the terms "dissolution cave" and "non-dissolution cave", in place of "solution cave" and "non-solution cave", and "pre-glacial" and "post-glacial" are meant as "pre-Wisconsinan" and "post-Wisconsinan", unless otherwise specified precisely. The reader should be aware that the scientific publication of the National Speleological Society changed its name from *Bulletin of the National Speleological Society* to *Journal of Cave and Karst Studies* with the start of the 1996 publishing year.

The authors gratefully acknowledge the many people who have worked in this region on finding, exploring, and explaining the caves and karst. Chuck Porter, the editor of the *Northeastern Caver*, must be lauded for his long years of effort, gathering cave and karst information from the region, and publishing it in an effective, complete, and continuous manner. He was of great assistance to the authors in collecting facts, images, and maps; he has also been an active cave explorer in the region for more than half a century. Dr. Arthur Palmer, and his wife Margaret have been the leading scientific influence for cavers both young (Cooper) and old (Mylroie) since the 1960s; they also gave permission to use many of their maps and images. Their insights, advice, and support were invaluable. Michael Chu and John Dunham provided exceptional assistance, especially access to their excellent image libraries of northeastern caves; Eric Cooper, Tom Feeney, and Alex Bartholomew assisted with images, and Michael Nardacci provided important comments on draft chapters. The late Bob Carroll was a significant participant in northeastern caving, especially remote marble and pseudokarst areas of the Adirondacks and New England; his fieldwork, commonly done solo, greatly expanded our knowledge of karst and pseudokarst in the northeastern United States. The Northeastern Cave Conservancy has helped maintain cave conservation and cave access in the region for decades; Bob Addis, Emily Davis, and Thom Engel in particular have been helpful to the authors. Joan Saxon Mylroie has been a loyal and effective field companion for more than 47 years. The list of all who contributed over the years would be quite long. The principal players have been cited and appear in the bibliography and we apologize in advance for anyone we have missed.

The *Northeastern Caver* remains the best single source of information on caves in the northeastern United States (Higham 2013a); that publication is now going electronic, which will make its reach even wider and more useful. The long scope of history about caves and karst development in the glaciated northeast makes it likely that the authors have missed important places and events, omitted significant contributions by others, and perhaps misinterpreted both facts and opinions. The project was done under tight time constraints and the authors ask forgiveness in advance for all errors, for which we assume full responsibility. As noted in Mylroie and Mylroie (2004, p. 92) "Ernst Kastning once said 'if all the caves in New York were laid end-to-end, they would be low, wet, muddy, and sparse in formations.' We all went and worked there anyway." And so we did.

References

Baker VR (1973) Geomorphology and hydrology of karst drainage basins and cave channel networks in east central New York. Water Resour Res 9:695–706

Baker VR (1976) Hydrogeology of a cavernous limestone terrane and the hydrochemical mechanisms of its formation, Mohawk River Basin, New York. Empire State Geogram 12:2–65

Cooper MP, Mylroie JE (2014) Post-glacial speleogenesis: verification of a hypothetical model, and the origins of maze caves in glaciated terrains. Cave and Karst Science 41(2):84–95

Dumont KA (1995) Karst hydrology and geomorphology of Barrack Zourie cave system, Schoharie County. Bulletin V of the New York Cave Survey, New York

Engel TD (2009) New York State. In: Palmer AN, Palmer MV (eds) Caves and Karst of the USA. National Speleological Society, Huntsville, Alabama, pp 27–32

Faulkner T (2008) The top-down, middle-outwards model of cave development in central Scandinavian marbles. Cave and Karst Science 34(1):3–16

Faulkner T (2009) Speleogenesis in New England marble caves. In: Proceedings of the Fifteenth International Speleological Congress, Kerrville, USA, vol 2, pp 855–862

Ford DC (1983) Effects of glaciations upon karst aquifers in Canada. J Hydrol 61(1):149–158

Ford DC (1987) Effects of glaciations and permafrost upon the development of karst in Canada. Earth Surf Process Landforms 12(5):507–512

Ford DC, Williams PD (2007) Karst hydrogeology and geomorphology. John Wiley & Sons, West Sussex, England

Higham S (2013a) The Northeastern Caver cumulative index: Volumes I–XLIV, 1969–2013. http://caves.org/region/nro/necaver_index.html. Accessed 11 Sept 2014

Jennings JN (1971) Karst. MIT Press, Cambridge

Kastning EH (1975) Cavern development in the Helderberg Plateau, East-Central New York. Bulletin I of the New York Cave Survey

Lauritzen S-E (1981) Glaciated karst in Norway. In: Proceedings, 8th International Speleological Congress, vol 2. Bowling Green, pp 410–411

Lauritzen S-E (2010) Grotter. Norges ukjente underverden (Caves. Norway's Unknown Underground). Tun Forlag, Oslo

Lauritzen S-E, Skoglund RØ (2013) Glacier ice-contact speleogenesis in marble stripe karst. In: Schroder JF (ed), Treatise on Geomorphology 6:363–396

Mylroie JE (1977) Speleogenesis and karst geomorphology of the Helderberg Plateau, Schoharie County, New York. Bulletin II of the New York Cave Survey

Mylroie JE (1984) Pleistocene climatic variation and cave development. Nor Geogr Tidsskr 38: 151–156

Mylroie JE, Mylroie JR (2004) Glaciated Karst: how the Helderberg Plateau revised the geologic perception. Northeast Geol Environ Sci 26(1–2):82–92

Palmer AN (1991) Origin and morphology of limestone caves. Geol Soc Am Bull 103:1–21

Palmer AN, Palmer MV (eds) (2009) Caves and karst of the USA. National Speleological Society, Huntsville, Alabama

Palmer MV (1976) Ground-water flow patterns in limestone solution conduits. Thesis, State University of New York at Oneonta

Perzan Z, Munroe JS, Amidon WH (2014) Investigation of last interglacial sediment in Weybridge Cave, Vermont, USA. In: Geological Society of America, Abstracts with Programs, vol 46(6), p 349

Porter C (2009) New England. In: Palmer AN, Palmer MV (eds), Caves and Karst of the USA. National Speleological Society, Huntsville, Alabama, pp 24–26

Quick PG (2010) Vermont caves: A geological and historical guide. National Speleological Society, Huntsville, Alabama

Waltham AC (1974) Limestone and caves of Northwest England. David and Charles, Newton Abbot, Great Britain

Waltham T, Lowe DJ (2013) Caves and Karst of the Yorkshire Dales. British Cave Research Association, Buxton

Weremeichik JM, Mylroie JE (2014) Glacial Lake Schoharie: An investigative study of glaciolacustrine lithofacies in caves, Helderberg Plateau, Central New York. J Cave Karst Stud 76(2):127–138

Wilderberger A, Preiswerk W (1997) Karst and Caves of Switzerland. Speleo Projects, Basel

Contents

Part II Cave and Karst Regions of the Northeastern US

Part I

Effects of Glaciation and Geology on Caves and Karst of the Northeastern US

Introduction

1

Abstract

The glaciated karst of the northeastern United States has added important knowledge about glaciation and speleogenesis, along with other well-studied continentally glaciated karst areas, such as Yorkshire (UK), and arctic Norway. In the northeastern US karst has developed on Precambrian marbles, Cambro-Ordovician marbles, highly folded and faulted Siluro-Devonian limestones, and flat-lying Cambrian to Devonian limestones. This structural progression allows the specific influence of glaciation to be better recognized. The area has been repeatedly glaciated in the Pleistocene, but only the last glaciation (Wisconsinan or MIS 2–4) is clearly recognized in the landscape. Glaciation's major effects were quarrying away of surficial caves and karst, derangement of drainage at the local and regional scale by both bedrock sculpture and sediment deposition, re-arrangement of insurgences and resurgences of caves, creating sediment occlusion and backflooding conditions within caves, and opening joints via isostatic rebound to influence ground water flow. The influence of proximal or overlying glacial lakes affects speleogenesis, water budgets, and cave sediment deposition. Glaciation acts during the ice advance, stillstand, and retreat phases, overprinting existing caves and creating conditions for new cave development. Glaciation, once thought to be detrimental to speleogenesis, is now recognized as an important contributor to speleogenesis and cave complexity. Research in the northeastern United States helped establish this new paradigm.

To understand the glaciated karst of New York and New England a general framework is needed of the types of karstic rocks, their extent, and their structure. Also needed are the timing and general effects of glaciation during the Pleistocene, and the effects of glaciation on karst; much of our understanding of this topic comes from the northeastern United States. This context is painted in broad strokes here, with fine details discussed through the scientific studies within the northeastern United States and elsewhere, the geologic history of the region, and through case studies of specific caves and cave systems presented later. The broad context allows comparisons of the karst terrains in the northeastern United States to other classically studied continentally glaciated karst terrains in Britain, Canada, and Norway, and also allows the contrast with alpine glaciated areas such as the Alps, the Rockies, and the Pyrenees.

M.P. Cooper and J.E. Mylroie, *Glaciation and Speleogenesis*,
Cave and Karst Systems of the World, DOI 10.1007/978-3-319-16534-9_1,

1.1 Karstic Rocks of the Northeastern United States

Karstic rocks in the northeastern United States include limestone, dolomite, and marble. By relative abundance of the large caves, New York has the greatest extent of karst (Tables 1.1 and 1.2). New York also has the largest variety of karst forming units and areas (Fig. 1.1a). The dissolutional caves in New York form in limestone, dolomite, and marble. In the New England states karstic caves are formed primarily in marble with some in limestones and dolomites (Fig. 1.1b). Pseudokarst caves are sometimes the predominant large cave systems in these states (Table 1.2). The region has a total area of 126,544 square miles (327,757 km^2); New York has an area of 54,550 square miles (141,299 km^2), New England 71,994 square miles (186,458 km^2), with Maine being 49 % of New England's total area. The preponderance of large caves and large karst areas in New York is a result of its large area, as well as its specific geology.

The northeastern United States contains several karst regions defined by their lithologic, stratigraphic, and structural components (Fig. 1.1). The units of these karst areas are Precambrian to Devonian in age, and vary in structure from flat-lying strata with a 1–2°

Table 1.1 The longest 25 limestone dissolution caves in the northeastern United States (from Higham 2013)[a,b,c]

Cave name	State	County	Length (ft)	Length (m)
		Limestone caves		
McFails	NY	Schoharie	37,528	11,439
Skull	NY	Albany	20,116	6131
Barrack Zourie	NY	Schoharie	16,718	5096
Glen Park Labyrinth	NY	Jefferson	14,000	4267
Thunder Hole	NY	Schoharie	10,356	3157
SCAG Maze	NY	Jefferson	10,000	3048
Mystery (Surprise)	NY	Sullivan	9974	3040
Howe Caverns	NY	Schoharie	8400	2560
Three Falls Complex	NY	Jefferson	7000	2114
Secret-Bensons	NY	Schoharie	6000	1829
Crayfish Water Labyrinth	NY	Jefferson	6000	1829
Onesquethaw-Jordans	NY	Albany	5500	1676
Clarksville	NY	Albany	4800	1463
Knox	NY	Albany	4170	1271
Schoharie Caverns	NY	Schoharie	4065	1239
Barytes-Wolferts	NY	Schoharie	3855	1175
Pompeys	NY	Ulster	3788	1155
Hailes	NY	Albany	3700	1128
Diddly	NY	Albany	3600	1097
Mangy Maze	NY	Jefferson	3300	1006
Gage Caverns (Balls Cave)	NY	Schoharie	3000	914
Single X	NY	Schoharie	2895	882
Chatter-Stone-Hole	NY	Albany	2850	869
Newell Street	NY	Jefferson	2600	792
Ain't No Catchment (ANC)	NY	Schoharie	2500	762
Knox-Fossil-Beckers	NY	Albany	2000	610
Hannacroix Maze	NY	Albany	2000	610

[a] At 2500 ft (762 m) long or longer
[b] Knox-Fossil-Beckers and Hannacroix Maze included as they are 2000 ft (610 m) long and mentioned in the text
[c] all Jefferson County caves (6) are maze caves

Table 1.2 The five longest marble and talus caves in the northeastern United States (Higham 2013)

Cave Name	State	County	Length (ft)	Length (m)
		Marble caves		
Aeolus Bat	VT	Bennington	3077	938
Merlins	NY	Columbia	2008	612
Natural Bridge	NY	Jefferson	2000	610
Burroughs	NY	Essex	2000	610
Vermonster	VT	(confidential)	2000	610
		Talus caves		
TSOD	NY	Essex	2400	732
MBDATHS	NH	Grafton	2100	640
Old TSOD	NY	Essex	2100	640
Scotts	NH	Grafton	2000	610
TSOD-Cyclops	NY	Essex	2000	610

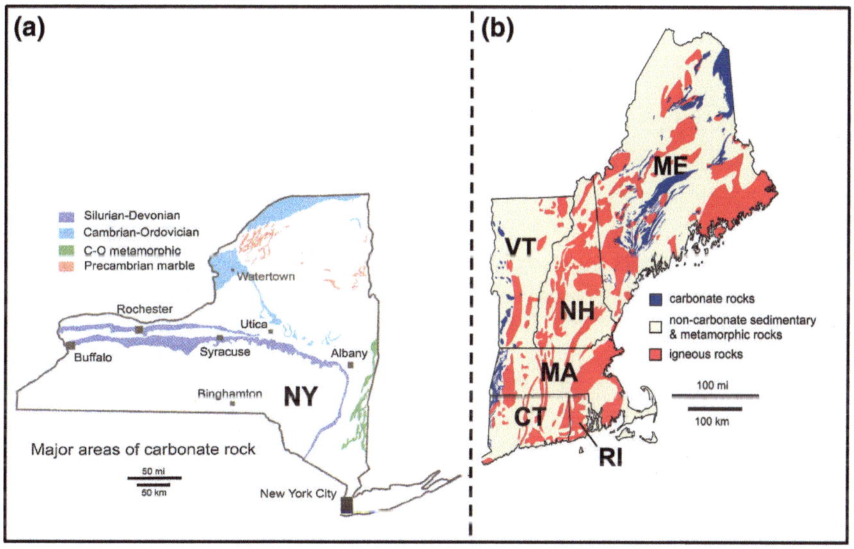

Fig. 1.1 Karst areas of New York and New England. **a** Map of New York showing outcrops of soluble rock (from Engel 2009). **b** Map of New England showing outcrops of soluble rock. Note that the western New England marble bands, especially in Massachusetts, spill over into eastern-most New York (from Porter 2009)

dip in the Helderberg Plateau, to highly metamorphosed, folded and faulted conditions such as in the Adirondacks. These units have formed over several tectonic periods and in different basins. The overall pattern for these units is marbles in the Adirondacks, marbles and limited limestones in New England and eastern New York, limestone in the northern lowlands and Mohawk Valley of New York, and limestone and dolomite along the rim of the dissected Alleghany Plateau in western and central New York. Breaking these down further and by age there are:

1. Precambrian marbles of the Adirondacks,
2. Precambrian, Cambrian, and Ordovician marbles of New England and eastern New York,
3. Cambro-Ordovician limestones of the northern New York and western Vermont lowlands and Mohawk Valley,
4. Siluro-Devonian limestones and dolomites in a relatively non-deformed east-west band in central-to-western New York, including the Helderberg Plateau,
5. and Siluro-Devonian limestones and dolomites in a structurally deformed north-south band in east-central New York, west of the Hudson River.

Early Devonian carbonates also exist within Maine, Massachusetts, and Vermont. Because of the geologic similarity of karst-forming rocks in New York east of the Hudson River to the karst-forming rocks of New

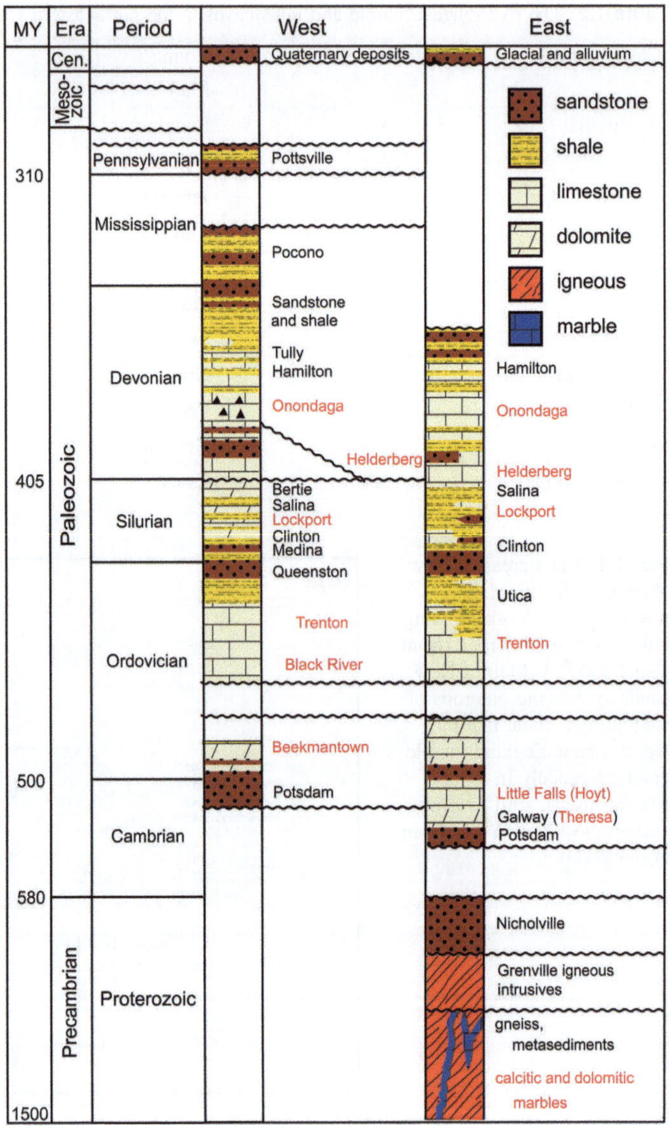

Fig. 1.2 Stratigraphic column for New York State. Karstic units in *red* (from Engel 2009)

England, those areas will be treated as a single region, ignoring the state political boundaries.

Stratigraphically most large cave systems found in the northeastern United States are within the Siluro-Devonian Helderberg Group, a group of limestone formations, particularly within the Manlius and Coeymans formations. Other stratigraphic units where caves form in New York are the Precambrian Grenville Marble, the Ordovician Black River and Trenton limestone groups, and the Devonian Onondaga and Becraft limestones (Fig. 1.2; Engel 2009). Sparse karstification has also occurred within the Cambro-Ordovician Chazy Limestone, Beekmantown Limestone and Little Falls Limestone, within the Silurian Cobleskill Limestone (even incising into the underlying Brayman calcareous shale) and Lockport Dolomite, the Siluro-Devonian Rondout Formation limestones and dolomite, and within the Devonian Tully Limestone. Geographically these units can be seen in Fig. 1.1.

Dissolutional cave development in New England and eastern New York is primarily limited to Cambro-Ordovician marbles of the Stockbridge Group (termed Wappingers Group in New York), and the Walloomsac Formation (Fig. 1.3; Plante 1990), though some cave development has taken place in limited outcrops of limestone in New England (Quick 2010). The

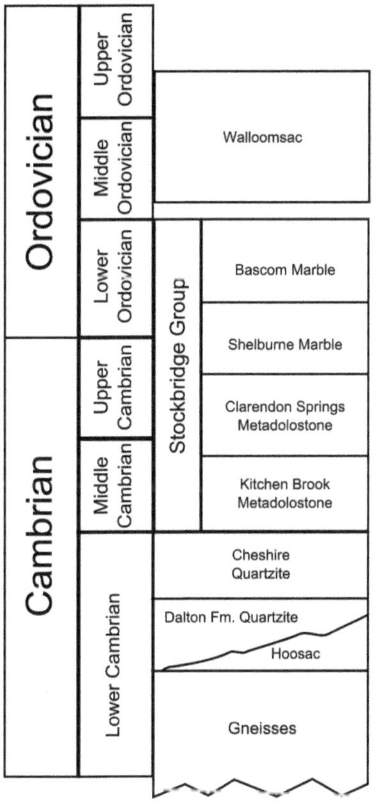

Fig. 1.3 Stratigraphic column of units in western Massachussettes and Vermont. Karstic rocks include units from the Stockbridge Group, and limestones within the Walloomsac Formation (Adapted from Plante 1990)

Sherman Marble in Vermont is Precambrian in age (Porter 2009). The majority of caves formed in the marbles of eastern New York and New England are located in the Berkshires of Massachusetts, as well as in Vermont. Dissolutional cave development has occurred more rarely in Connecticut and is almost absent in Rhode Island. New Hampshire lacks dissolutional caves. Maine has soluble rock outcrops of some extent, but little is known about the karst there. While there are fewer known dissolutional caves in New England compared to other karst regions around the US, there are abundant pseudokarst caves in these states (e.g. Carroll 1997), which are greatly influenced by glaciation (Chap. 4). These pseudokarst caves can be found in younger rocks than those that form dissolutional karst caves, including Mesozoic lavas and granites. Constructional travertine and tufa caves also exist in New York: these caves being not tied to glaciation, are most likely Holocene and therefore post-glacial in origin (Fig. 1.4).

The metamorphism and structural development of these karstic rocks is due to several orogenic events: the Grenville Orogeny of the Precambrian, the Taconic Orogeny of the Cambrian and Ordovician, the Acadian Orogeny of the Devonian, and the Alleghanian Orogeny of the Mississippian-to-Permian. Pseudokarst caves have developed in rocks influenced by Triassic rifting in the Connecticut River valley, and post-Paleozoic intrusions in New England. The Adirondack

Fig. 1.4 Calcareous tufa deposition and tufa caves in New York. **a** Calcareous tufa in the stream bed draining Tufa Cave, Schoharie County, New York, the low-flow resurgence for Schoharie Caverns (Chap. 5). This tufa rests directly upon glacial till, and so while voluminous, is post-glacial in origin. **b** Vanhornsville Tufa Caves, Schoharie County—Herkimer County line, New York. This deposit is post-glacial, yet contains humanly negotiable passages. It is a constructional cave (*sensu* Mylroie and Mylroie 2013, their Fig. 1.4) in the sense that the void was already there, and tufa deposited around it, as opposed to the void being dissolved out of the tufa

Mountains are uplifting, both from glacial rebound and also from active doming, with contemporary minor earthquakes common (Isachsen et al. 2000).

1.2 Glaciation

In higher latitudes and altitudes during the Pleistocene, continental and alpine glaciation was the primary geomorphic process influencing landscapes and their evolution, and the landforms glaciers leave behind continue to influence geomorphic development during interglacial periods. While continental glacial cycles are present in the Quaternary, they are rare in geologic time, with other major occurrences in the Precambrian, Ordovician, and Late Carboniferous/Early Permian (Hambrey and Harland 1981). The occurrence of glaciation requires "just right" conditions of orbital cyclicity, atmospheric composition, position of continents, as well as other factors. Continental position relative to high latitudes is the most likely reason for major glaciation being a rare event when viewed across the span of the Phanerozoic. Continental glaciation within the current ice age of the Quaternary has predominantly influenced the high northern latitudes as the bulk of continents are in the northern hemisphere, though ice sheets have influenced, and still exist on Antarctica. Alpine glaciation has taken place on all continents and all latitudes at high altitudes over a wider range of geologic time.

The current ice age can be broken into multiple glacial and interglacial periods, where glacial periods have established ice sheets that advanced towards lower latitudes and elevations, and interglacial periods are times of warmer temperatures and continental ice sheets existing only at very high latitudes. The current interglacial is the Holocene, and the most recent glacial period has a variety of names depending on geography and individual ice sheets. These names include the Wisconsinan for eastern North America, Devensian in Britain, Midlandian in Ireland, and the Weichselian in Scandinavia and northern Europe. Other names exist for continental glaciations as well as for alpine glaciations such as the Würm in the Alps (Gibbard and Clark 2011). The previous interglacial also has varying names depending on geography including Eemian in northern Europe, and Ipswichian in Britain. The glacial period prior to the Wisconsinan includes the Illinoian in North America, and the Wolstonian in Britain. The glacial periods of the current ice age have spans of 60–100 ka, while interglacial periods last 10–20 ka.

These glacial and interglacial periods line up with the Marine Isotope Stages (MIS) (Fig. 1.5), with the current interglacial being MIS1, the recent glacial being MIS2–4, and the previous interglacial being MIS5 (subdivided into relative sea-level highstands labeled MIS substages 5a, 5c and 5e, with 5e being the longest and 6 m higher than present; modest lowstands occurred at MIS substages 5b and 5d). The MIS 3 highstand has been the subject of much debate as to its actual sea-level elevation, but is currently believed to be below modern (however, see Mylroie and Carew 1988). Evidence exists to suggest that ice withdrawal during MIS 3 was sufficient in New York to leave a signal in cave stalagmites (Lauritzen and Mylroie 2000), as discussed later. The use of the MIS system allows finer resolution of warm and cold periods, and also allows timing further back than landforms allow, as subsequent glaciations reset surficial landscapes and obscure prior events.

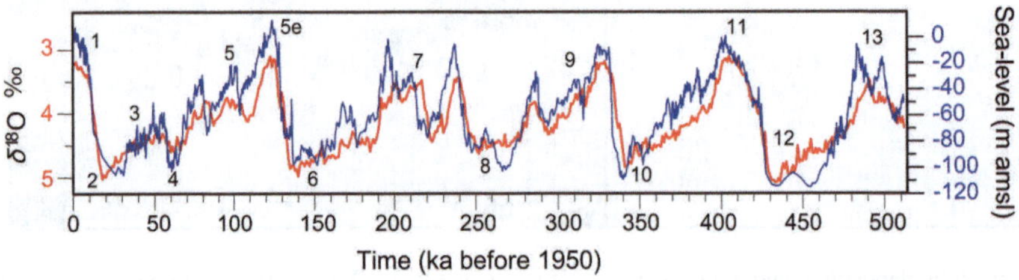

Fig. 1.5 Marine Isotope Stage and sea level curves for the last 500 ka. Stable oxygen isotope values from deep-sea cores are an excellent proxy for ice volume, and thus glacial/interglacial cycles. The last glacial cycle (Wisconsinan) corresponds to MIS 2–4. The last sea-level highstand (MIS 5e) was 6 m higher than present. (Modified from Badertscher et al. 2011 with isotope data from Lisiecki and Raymo 2005, and sea-level data from Siddall et al. 2003 and Rohling et al. 2010)

Glaciation in the Pleistocene initiated about 2.6 Ma, and underwent a major shift from being forced by 41 kyr orbital cycles to 100 kyr orbital cycles about 0.8 Ma (Lauritzen and Skoglund 2013). As each glaciation erodes and modifies many of the landforms produced by the prior glaciation, few landscapes as seen today can be attributed to ice behavior prior to 0.8 Ma. The traditional and obsolete North American convention of four major glaciations (Nebraskan, Kansan, Illinoian, and Wisconsinan) was based on survival of terminal lobes of glacial sediment from each event. The development of the MIS system using sea-floor data allows a fuller understanding of glacial episodes going back to the start of the Pleistocene and beyond. Data from Norway (Lauritzen and Skoglund 2013), Great Britain (Waltham and Lowe 2013), and New York (Lauritzen and Mylroie 2000) indicate that dissolutional caves can survive multiple glaciations and may provide data from the terrestrial environment that compliments or improves upon that from the ocean floor.

1.2.1 Glacial Effects

Through the glacial cycle there are numerous effects of glaciation that reset the surficial topography, geology, and hydrology, as well as effects on water chemistry from permafrost and removal of soils. These effects include action from the ice of the glacier itself, as well as the meltwaters, sediment outwash, and ice-dammed lakes that accompany ice sheets. Broadly the effects can be classified as thermal, mechanical, chemical, and hydrological, with these effects being commonly linked with each other. Each of these effects operates at a variety of scales that can be seen in maps and aerial views. Effects are not limited to glaciated areas but to the entire planet from fluctuating sea levels and shifts in atmospheric circulation patterns.

Mechanical effects include those from ice-sheet movement and meltwaters eroding, transporting, and depositing sediments. On the large-scale, ice lobes erode valleys, scour plains and plateaus, and create large lake basins. Large-scale removal of material occurs due to mechanical plucking and abrasion that occurs basally and laterally by the ice sheet. Erosion by ice can create over-steepened U-shaped valleys that can be later inhabited by underfit streams (Fig. 1.6), and can also create steep escarpments (Fig. 1.7). Large-scale erosion can also be due to meltwaters and the failing of ice dams that hold ice-dammed lakes (de Simone et al. 2008). Material eroded by glaciers can subsequently be transported and deposited and can create large-scale features such as Long Island, New York (Fig. 1.8). This material is deposited in the form of unsorted sediment named till, and large-scale deposition can occur in moraine landforms, particularly terminal moraines, and can also bury pre-glacial

Fig. 1.6 An underfit stream at Sally Gap, County Wicklow, Ireland. This small stream flows through a much larger, broad, U-Shaped valley. (*Photo* courtesy E. Cooper)

Fig. 1.7 Thacher Park, Albany County, New York. This park was established to take advantage of the outstanding vista and significant geology found here (see also Chap. 5, Fig. 5.3). **a** View northeast, overlooking the Mohawk-Hudson lowland. The cliff is made up of the Coeymans Limestone overlying the Manlius Limestone of the Helderberg Group, which form a resistant surface that creates high scarps as seen here, and a plateau surface extending to the south. The lower exposed slope is in the Schenectady Formation, a series of Ordovician shales and siltstones with less strength. Glaciers have exploited this difference in strength to accentuate the relief seen here. **b** Close up of the cliff at Thacher Park, with the massive Coeymans Limestone forming the upper part of the cliff, and the thin-bedded Manlius Limestone making up the lower half. Note the overhanging nature of the resistant Coeymans Limestone. Person in *circle* for scale

Fig. 1.8 Landsat image of Long Island, New York. This island is composed of the terminal moraine for the Wisconsinan glaciation, and is 190 km (118 miles) long. (Image from NASA Landsat imagery)

river valleys (Palmer 1976). Silt-sized eroded grains derived from glacial erosion can be deposited hundreds or thousands of kilometers away by wind as loess deposits.

Smaller scale mechanical effects mirror the larger with erosion, transport and deposition. Plucking and abrasion form numerous small depressions that can be aligned with local ice-movement direction, and when the geology allows ice-aligned ridges can be formed (Fig. 1.9). Plucking and abrasion also create landforms such as roche moutonnée (also called whalebacks, Fig. 1.10a). Deposition from meltwaters around melting ice blocks can also create water-filled depressions called kettles (Fig. 1.10b). Other small-scale erosional features include glacial striations (Fig. 1.10c) formed due to abrasion on bare rock surfaces, also aligned with ice-direction. Depositional features include elongated hills of till called drumlins (Fig. 1.10d), mounds called kames, and sinuous outwash deposits called eskers. Certain features can group together such as drumlins into drumlin fields, and kames and kettles in kame and kettle topography. Another depositional product is an erratic, a boulder of different lithology from the underlying terrain as a result of ice transport. These erratics can be moved hundreds of kilometers and can range from cobble

Fig. 1.9 Exaggerated (6.4 times) DEM of the area north of Joralemon Park, Albany County, New York. Seen here are north-south oriented ridges, valleys, and depressions. These ridges are a series of anticlines and synclines, and are a continuation of the Ridge and Valley province. The axes of these folds run north-south. North-south ice movement over this region further enhanced the exposure of these folds

sized to boulders as large as a house (Fig. 1.10e). When erratics are left behind on soluble rocks, they form karrentisch (literally, "karst table"; Ford and Willams 2007), as dissolutional denudation does not occur under the protective cap provided by the erratic (Fig. 1.11a). From karrentisch, knowing the age of the emplacement of the erratic, estimates can be made of carbonate rock denudation. As glaciers commonly planate the surfaces they move over, insoluble components of a carbonate rock may weather out in positive relief, again providing an estimate of post-glacial denudation rates (Fig. 1.11b).

Chemical effects of glaciation are due to lower temperatures, establishment of permafrost, and removal of soils. Lower temperatures decrease the

Fig. 1.10 Photographs of small-scale glacial landforms. **a** Roche moutonée in Yorkshire, England. Roche moutonées can indicate ice-flow direction: the lee (steep) side pointing in the direction of flow. Camera direction is to the west, indicating a southerly ice-flow direction. **b** Aerial photograph of kettles in Chugach National Forest, Alaska. Largest kettle is ~150 ft (~45 m) in diameter. **c** Striations on a schist outcrop on the Eastern Summit of the Berkshires. Camera direction and striation direction is to the south, indicating southerly ice-flow. **d** Drumlin at Cranes Beach, Massachusetts. People for scale in *red circle*. **e** Erratic in Glomdal, Norway. (Photos b and d from USGS Glossary of Glacier Terminology)

Fig. 1.11 Denudation rates of glaciated karst. **a** karrentisch ("karst table") Glomdal, Norway, a granite boulder on Cambro-Ordovician marble, displaying ~15 cm of denudation since ice retreat about 8 ka. **b** Quartz vein in Cambro-Ordovician marble in Glomdal Norway, rising >10 cm above the surrounding marble, another indicator of marble denudation since ice retreat 8 ka

solubility of most minerals, limiting chemical weathering. Counter-intuitively, calcium carbonate is more soluble at lower temperatures; however the actual effect is quite complex (Lauritzen and Skoglund 2013). Permafrost is established at high latitudes and along the margins of ice-sheets, where it can stretch for 100+ km (60+ miles) (French and Millar 2014). Permafrost decreases organic carbon production and reduces soil development. The freezing of organic matter also reduces the release of soil gasses into the atmosphere. Subsequent thawing of this organic matter after deglaciation can release organic carbon back into the atmosphere (Hodgkins et al. 2014). Permafrost also decreases the amount of dissolved species that enter groundwater. Removal of established soils decreases available organic matter.

Hydrologic effects of glaciation include physical and chemical hydrogeologic effects, as well as surficial hydrologic effects. These effects span worldwide as available water is tied up in ice-sheets, lowering eustatic sea level (Fig. 1.5), and creating pluvial environments (wetter, cooler areas that are dry during subsequent interglacials). During glacial times groundwater level is increased in areas under ice-sheets, as pressures from the overlying ice increases head (Mylroie 1984; Lauritzen and Skoglund 2013). Under pro-glacial and ice-dammed lakes, base level increases to the surface of the lakes. Groundwater chemistry under ice-sheets and permafrost decreases concentrations of chemical species from soils, as they cannot be replenished. Lauritzen and Skoglund (2013) describe the "carwash effect" of glaciation on karst areas, in which the landscape, like the car in a car wash, is stationary while the ice moves back and forth across it. This model means that the karst area can expect pre-ice, ablation ice, and accumulation ice environments as the glacial ice comes, to be repeated when the ice goes. They modeled both the physical hydrology of this process, and the chemical shifts that occur at each stage.

Surficial hydrologic effects during glacial times include the establishment of meltwater streams, subglacial lakes, proglacial lakes, and ice-dammed lakes. During deglaciation meltwater discharge dramatically increases, and can route through non-glaciated areas. Ice-dammed lakes can be established during deglaciation, and their failure can also route large discharges through valleys, establishing large, broad valley cross-sections that currently contain underfit streams and rivers (de Simone et al. 2008). Moraine-dammed lakes can survive after deglaciation and exist during interglacials.

These effects have dramatic impacts on the post-glacial landscape, with the surface topography and hydrology reset. Neither surface hydrology nor groundwater hydrology fully returns to pre-glacial patterns. Surface hydrology is guided by the new landforms, with lakes forming in glacial depressions, and surface streams being guided by drumlins, moraines, and other depositional landforms in a pattern called deranged drainage (Fig. 1.12; Twidale 2004). Swamps/marshes can replace deglacial lakes in depressions as small surface streams poorly drain the depressions, and sediment collects. Rivers in the post-

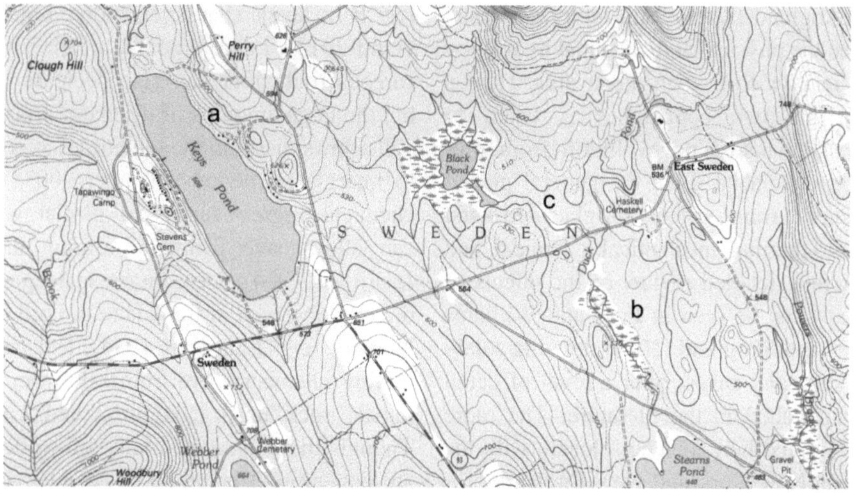

Fig. 1.12 Section of the North Waterford Quadrangle of Maine, showing deranged drainage typical of glaciated regions. This drainage style, when caused by glaciation, is characterized by numerous, glacially carved depressions filled with lakes (**a**) when drained efficiently, or swamps/marshes (**b**) when drained poorly. These lakes and marshes are connected by small streams (**c**). This drainage style typically ignores underlying geology, and is instead guided by erosional and depositional glacial landforms, such as those seen in Fig. 1.10. (Section from: USGS North Waterford Quadrangle 1995)

glacial landscape can have entirely new courses as the ice sheet erodes pre-glacial drainage divides or glacial sediment infills earlier stream courses. Groundwater does not usually return to a previous, pre-glacial water table, as base level is lowered by erosion, or elevated by infilling of pre-glacial valleys (Palmer 1976). The deglacial period is also marked by changes in precipitation and evapotranspiration. After glacial retreat worldwide eustatic sea level is also elevated. Of important note is that the glacial landforms, surficial hydrology, and subsurficial hydrology of previous interglacials are mostly removed, and further reset upon subsequent glaciation. As will be seen later, caves, because they are *within* the rock and not *on* the rock, can survive multiple glaciations and help re-establish pre-glacial hydrology.

Isostatic effects also take place after deglaciation, with uplift of previously depressed crust. This uplift can produce neotectonic earthquakes, and can also mechanically enlarge joints and other fractures (Harland 1957). In addition, basins change due to crustal tilting, and marine water bodies become replaced by freshwater lakes as the crust rebounds above sea level (e.g. Lake Champlain on the Vermont/New York border, Chap. 8).

1.2.2 Glacial Effects on Caves and Karst

As karst is a function of recharge, chemistry, and geology, glaciation has an extreme impact on karst. In glaciated terrains entire caves can be removed by glaciation, or be buried and preserved by infilling with sediment. Worldwide, sea-level variation can change base level and change the position of active cave development, and may entirely change the mode of cave development by allowing more recharge (Vacher and Mylroie 2002). During pluvial times areas that are currently arid receive larger recharge that enhanced cave development.

Glaciated karst areas are affected in various ways by glaciation, and the effects can be discussed in a variety of contexts, including by *effect categories* (Ford 1983), and by *glacial ice position* (Mylroie 1977, 1984; Lauritzen and Skoglund 2013); Ford (1983), and Ford and Williams (2007, their Table 10.2) categorize effects based on the state of the glaciated karst, including being protected by burial, or destroyed by erosion. Mylroie (1977, 1984) classified effects on position of ice, where ice was advancing and in proximity including periglacial effects, where ice was covering the karst, and where ice was retreating.

Lauritzen and Skoglund (2013) with their "carwash" model, quantified the physical and chemical hydrology of ice position relative to a given karst area. Here a combination of the two approaches is taken to understand the karst landscape evolution from pre-glacial times to post-glacial. Like the general glacial effects on glaciated terrains, effects on glaciated karst can be broken into thermal, mechanical, chemical, and hydrological. A full, detailed review of the effects of glaciation on karst can be seen in Ford and Williams (2007), Lauritzen and Skoglund (2013) drawing solely on Norwegian karst and caves, offer a level of detail and example unmatched in the literature. The more specific view presented here matches closely with the conditions seen in New York and New England.

In the early stages of ice advance periglacial environments exist, where permafrost can be established. The implications of permafrost on karst are reduction in P_{CO_2} reducing aggressiveness of the water, and blocking recharge into karst aquifers both from sinking streams and diffuse recharge (Ford 1977). Limiting recharge by permafrost can drop the water table, thus changing previously phreatic conditions into vadose, though speleothem growth in these vadose conditions may be shut off by lack of calcite in solution (Lauriol et al. 1997) and the lack of available CO_2 to degas. Permafrost also mechanically erodes by freeze-thaw weathering and can create pseudokarst forms such as depressions, dry valleys, and caves (Halliday 2007), though during the advancing-ice stage these features are likely to be removed.

As ice sheets advance over karst terrains plucking and abrasion can remove surficial karst and shallow caves, though limestones are preferentially left by glaciation, as many clastic sedimentary rocks such as shales are mechanically weaker. Early thoughts were that this advancing ice would entirely remove caves by plucking or abrasion, or would crush caves under the mass of the ice (Jennings 1971), though it has been shown that deeper, pre-glacial caves survived glacial erosion in a variety of continentally glaciated areas (e.g. Great Britain, Waltham 1974; New York, Mylroie 1977; Norway, Lauritzen 1981). During the ice-contact phase recharge can occur from meltwater, and from the internal plumbing of the ice sheet routing water to the base of the glacier (Fig. 1.13). As the solubility of calcite increases as temperatures decrease (Adams and Swinnerton 1937; White 1988), dissolution can still take place, though it is diminished due to low P_{CO_2} levels (Faulkner 2006a; Lauritzen and Skoglund 2013). Large amounts of glacially derived calcareous rock flour can also decrease aggressiveness, but the large amounts of discharge from glacial waters still allows dissolution, sometimes at comparable rates to interglacial periods (Tranter 2003). Recharge and chemical aggressiveness of glacial waters therefore can allow cave formation during ice-contact (e.g. Lauritzen 1984). Ice-contact speleogenesis has been directly observed in Canada (e.g. Ford et al. 1983) and Norway (e.g. Lauritzen 1984), and can be aided by mixing chemistry with minerals from adjacent, non-karstic rock units (Skoglund and Lauritzen 2011). Though there is some body of work on subglacial speleogenesis (including a chapter in Treatise on Geomorphology: Lauritzen and Skoglund 2013 there is debate in the literature (e.g. Faulkner 2009) on its occurrence, as subglacial flow routes are invoked and it may be the case that calcite saturation proceeds quickly in these routes. Subglacial $CaCO_3$ precipitation may occur as well, such as crusts deposited from pressurized melt waters (Ford and Williams 2007).

During deglaciation material is deposited as till and outwash, plugging cave entrances and mantling karstic rocks. Plugging and mantling can be considered a protective effect (Ford 1983), preserving pre-glacial caves, limiting surface dissolution, and preserving surficial forms on karstic rocks such as glacial striations (Chap. 5, Fig. 5.11). Glacial sediment deposits within caves can also stimulate passage enlargement, as hydraulic head is raised and floodwater mazes form as a result (Chap. 5, Fig. 5.9; Palmer 1975). Deglacial seismicity and rebound allows mechanical expansion of joints (Harland 1957; Faulkner 2006b), also aiding in the rapid formation of passages in post-glacial caves by decreasing the need for long breakthrough times. Unique cave deposit lithofacies are generated by glaciation including diamicton facies, where ice directly forces sediment into the cave (Farrant and Smart 2011), and glaciolacustrine facies from the presence of glacial lakes over a cave's aerial footprint (Weremeichik and Mylroie 2014). During deglacials ice-dammed lakes can form, greatly raising hydraulic head and reducing breakthrough times dramatically (Faulkner 2008), especially in conjunction with mechanically enlarged joints. This style of speleogenesis can produce rapidly forming caves, though like subglacial cave development, there is active debate on the occurrence of this form of speleogenesis (Lauritzen et al. 2009; Faulkner

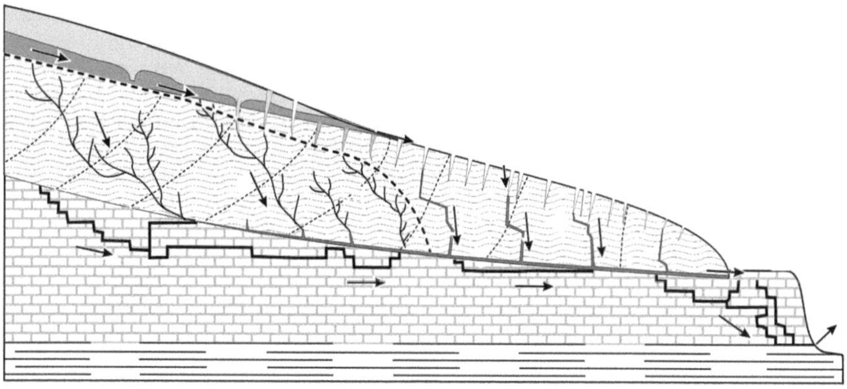

Fig. 1.13 Routing of water flow through and under ice, with input into karst systems. Here water flows through moulins, down to the base of the ice sheet where it can enter the limestone, where it is guided to an ultimate dissolutional or hydrologic base level. Thick dashed line is the potentiometric surface. (Adapted from Mylroie 1984 and Lauritzen and Skoglund 2013)

2009; Lauritzen and Skoglund 2013) as the deglacial time period is short compared to glacial and interglacial periods, and neotectonic effects may be overstated (Bungum et al. 2010; Lauritzen and Skoglund 2013). Deglacials also shortly return glaciated karst terrains to periglacial environments, with a return of permafrost and the effects of the permafrost. Depending on position relative to the retreating ice sheet, deglacial permafrost forms may persist if no glacial re-advances override them.

The landforms left by glaciers have a dramatic impact on the interglacial karst landscape. Pre-glacial karst forms can persist into the post-glacial time period if protected by sediments, or if deep enough to be unaffected by glacial erosion, and yield speleothem dates up to the limit of dating techniques ($\sim$600 ka, White 2004; Ford and Williams 2007; 700 ka, Lauritizen and Skoglund 2013) indicating their survival through multiple glacial and interglacial periods. Pre-glacial cave systems can have old entrances occluded or removed, and new entrances can form by removal of bedrock. Shallow caves and pre-glacial karst forms from the previous interglacial are removed unless protected at high elevations, or are outside the footprint of the previous advance. Barren karst formed by glacial soil removal, combined with mechanical joint expansion can produce limestone pavements (Chap. 5, Fig. 5.12), and other dramatic forms of karren (especially in alpine glaciated karst). Pre-glacial caves may reenter the flow regimes of the interglacial period (Kastning 1975; Baker 1976; Mylroie 1977; Dumont 1995) fed by post-glacial infeeders (Mylroie and Carew 1987), and adjust to the post-glacial base level (Palmer 1976). Post-glacial karst forms are congruent with the current deranged drainage and follow flow routes off of and into glacial landforms (Mylroie and Carew 1987; Cooper and Mylroie 2014). These post-glacial forms can be infeeders into pre-glacial systems, or entirely new, post-glacial caves. The deranged drainage can produce conditions to allow rapid speleogenesis with lakes creating high hydraulic gradients, and swamps formed in poorly-drained glacial depressions can contain low pH, high P_{CO_2}, and organic acids, greatly increasing dissolution rates and creating high surface denudation rates (Allred 2004), as well as cave wall-retreat rates (Cooper and Mylroie 2014).

Glaciation was originally thought to be detrimental to the survival of cave and karst features, but it is now recognized that karst not only survives glaciation, but in many respects, can be enhanced by glaciation (Mylroie and Mylroie 2004). Pseudokarst, on the other hand, does not survive glaciation, but the glacial process, as will be seen in detail later (Chap. 4), greatly enhances pseudokarst cave production. Glaciation is an immense geological perturbation, and its effects in New York and New England were dramatic. Cave and karst features responded to that glacial impact in significant and interesting ways, as this book will attempt to demonstrate.

References

Adams CS, Swinnerton AC (1937) The solubility of limestone. Trans Am Geophys Union 11:504–508

Allred K (2004) Some carbonate erosion rates of southeast Alaska. J Cave Karst Stud 66(3):89–97

Badertscher S, Fleitmann D, Cheng H et al (2011) Pleistocene water intrusions from the Mediterranean and Caspian seas into the Black Sea. Nat Geosci. doi:10.1038/NGEO1106

Baker VR (1976) Hydrogeology of a cavernous limestone terrane and the hydrochemical mechanisms of its formation, Mohawk River Basin, New York. Empire State Geogram 12:2–65

Bungum H, Olesen O, Pascal C et al (2010) To what extent is the present seismicity of Norway driven by post-glacial rebound? J Geol Soc 167(2):373–384

Carroll RW (1997) Reassessment of talus caves and snow tubes. Northeast Caver 28(3):92–93

Cooper MP, Mylroie JE (2014) Post-glacial speleogenesis: verification of a hypothetical model, and the origins of maze caves in glaciated terrains. Cave and Karst Sci 41(2):84–95

de Simone DJ, Wall GR, Miller NG (2008) Glacial geology of the northern Hudson through the southern champlain lowlands. Guidebook to Field Trips for the 71st Annual Reunion, Northeastern Friends of the Pleistocene

Dumont KA (1995) Karst hydrology and geomorphology of Barrack Zourie cave system, Schoharie County. Bulletin V of the New York Cave Survey, New York

Engel TD (2009) New York State. In: Palmer AN, Palmer MV (eds) Caves and karst of the USA. National Speleological Society, Huntsville, pp 27–32

Farrant AR, Smart PL (2011) Role of sediment in speleogenesis; sedimentation and paragenesis. Geomorph 134:79–93

Faulkner T (2006a) Limestone dissolution in phreatic conditions at maximum rates and in pure, cold, water. Cave Karst Sci 33(11):11–20

Faulkner T (2006b) Tectonic inception in caledonide marbles. Acta Carsologica 35(1):7–21

Faulkner T (2008) The top-down, middle-outwards model of cave development in central Scandinavian marbles. Cave Karst Sci 34(1):3–16

Faulkner T (2009) Reply to "Marble caves and self-declared paradigms". NSS News

Ford DC (1977) Karst and glaciation in Canada. In: Proceedings, 7th international speleological congress. pp 188–189

Ford DC, Williams PD (2007) Karst hydrogeology and geomorphology. Wiley, West Sussex

Ford DC, Smart PL, Ewers RO (1983) The physiography and speleogenesis of Castle Guard Cave, Columbia Icefields, Alberta, Canada. Arct Alp Res 15:437–450

French HM, Millar SWS (2014) Permafrost at the time of the last glacial maximum (LGM) in North America. Boreas 43:667–677

Gibbard PL, Clark CD (2011) Pleistocene glaciation limits in Great Britain. In: Ehlers J, Gibbard PL, Hughes PD (eds) Quaternary glaciations—extent and chronology—a closer look. Developments in Quaternary Science, vol 15. Elsevier, Amsterdam, pp 75–93

Halliday WR (2007) Pseudokarst in the 21st century. J Cave Karst Stud 69(1):103–113

Hambrey MJ, Hardland WB (eds) (1981) Earth's pre-pleistocene glacial record. Cambridge University Press, Cambridge

Harland W (1957) Exfoliation joints and ice action. J Glaciol 3 (21):8–10

Higham S (2013) Northeast long caves, northeast deep caves. Northeast Caver 44(2):44–45

Hodgkins SB, Tfaily MM, McCalley et al (2014) Changes in peat chemistry associated with permafrost thaw increase greenhouse gas production. PNAS 111(16):5819–5824

Isachsen YW, Landing E, Lauber JM, et al. (2000) Geology of New York: A simplified account, Second Edition. New York State Museum, Albany, New York

Jennings JN (1971) Karst. MIT Press, Cambridge

Kastning EH (1975) Cavern development in the Helderberg plateau, East-Central New York. Bulletin I of the New York Cave Survey

Lauriol B, Ford DC, Cinq-Mars J, Morris WA (1997) The chronology of speleothem deposition in northern Yukon and its relationship to permafrost. Can J Earth Sci 34:902–911

Lauritzen S-E (1981) Glaciated karst in Norway. In: Proceedings, 8th. international speleological congress, vol 2. Bowling Green, pp 410–411

Lauritzen S-E (1984) Evidence of subglacial karstification in Glomdal, Svartisen, Norway. Nor Geogr Tidsskr 38:169–170

Lauritzen S-E, Mylroie JE (2000) Results of a speleothem U/Th dating reconnaissance from the Helderberg Plateau, New York. J Cave Karst Stud 62(1):20–26

Lauritzen S-E, Skoglund RØ (2013) Glacier ice-contact speleogenesis in marble stripe karst. In: Schroder JF (ed), Treatise on Geomorphol 6:363–396

Lauritzen S-E, Holbye U, Solbakk T (2009) Marble caves and self-declared paradigms. NSS News

Lisiecki LE, Raymo ME (2005) A Pliocene-Pleistocene stack of 57 globally distributed benthic δ18O records. Paleoceanography. doi:10.1029/2004PA001071

Mylroie JE (1977) Speleogenesis and karst geomorphology of the Helderberg Plateau, Schoharie County, New York. Bulletin II of the New York Cave Survey

Mylroie JE (1984) Pleistocene climatic variation and cave development. Nor Geogr Tidsskr 38:151–156

Mylroie JE, Carew J (1987) Field evidence for the minimum time for speleogenesis. Natl Speleol Soc Bull 49(2):67–72

Mylroie JE, Carew JL (1988) Solution conduits as indicators of late Quaternary sea level position. Quatern Sci Rev 7:55–64

Mylroie JE, Mylroie JR (2004) Glaciated karst: how the Helderberg plateau revised the geologic perception. Northeast Geol Environ Sci 26(1–2):82–92

Mylroie JE, Mylroie JR (2013) Pseudokarst caves in the littoral environment. In: Lace MJ, Mylroie JE (eds) Coastal karst landforms. Coastal research library 5. Springer, Dordrecht, pp 3–14

Palmer AN (1975) The origin of maze caves. Natl Speleol Soc Bull 34(3):56–76

Palmer MV (1976) Ground-water flow patterns in limestone solution conduits. Thesis, State University of New York at Oneonta

Plante AR (1990) Berkshire caves and karsts, Berkshire geology and karst landforms, Part-I: history, stratigraphy, and structure. Northeast Caver 21(2):48–56

Porter C (2009) New England. In: Palmer AN, Palmer MV (eds) Caves and karst of the USA. National Speleological Society, Huntsville, pp 24–26

Quick PG (2010) Vermont caves: a geological and historical guide. National Speleological Society, Huntsville

Rohling EJ, Braun K, Grant K, Kucera M, Roberts AP, Siddall M, Trommer G (2010) Comparison between Holocene and Marine Isotope Stage 11 sea level histories. Earth Planet Sci Lett 291:97–105

Siddall M, Rohling EJ, Almogi-Labin A, Hemleben Ch, Meischner D, Schmelzer I, Smeed DA (2003) Sea-level fluctuations during the last glacial cycle. Nature 423:853–858

Skoglund RØ, Lauritzen S-E (2011) Subglacial maze origin in low-dip marble stripe karst: examples from Norway. J Cave Karst Stud 73(1):31–43

Tranter M (2003) Geochemical weathering in glacial and proglacial environments. In: Drever JI (ed) Treatise on geochemistry. Elsevier, Amsterdam, pp 189–205

Twidale CR (2004) River patterns and their meaning. Earth Sci Rev 67:159–218

Vacher HL, Mylroie JE (2002) Eogenetic karst from the perspective of an equivalent porous medium. Carbonates Evaporites 17(2):182–196

Waltham AC (1974) Limestone and caves of Northwest England. David and Charles, Newton Abbot

Waltham T, Lowe DJ (2013) Caves and karst of the Yorkshire dales. Br Cave Res Assoc, Buxton

Weremeichik JM, Mylroie JE (2014) Glacial Lake Schoharie: An investigative study of glaciolacustrine lithofacies in caves, Helderberg Plateau, Central New York. J Cave Karst Stud 76(2):127–138

White WB (1988) Geomorphology and hydrology of karst terrains. Oxford University Press, Oxford

White WB (2004) Paleoclimate records from speleothems in limestone caves. In: Sasowsky ID, Mylroie JE (eds) Studies of cave sediments. Kluwer Academic/Plenum Publishers, New York, pp 135–175

Geologic and Geomorphic History of New York and New England

2

Abstract

The geology of the northeastern Unites States begins with the late Mesoproterozoic, over 1 billion years ago, when the Grenville suite of metamorphics (including the Grenville Marble) of the Adirondack Mountains was formed during the assembly of the supercontinent of Rodinia. Cambrian and early Ordovician carbonates, along with clastics, were deposited in continental shelf conditions until the mid-Ordovician Taconic Orogeny, when those carbonates were metamorphosed to marbles in eastern New York, southwestern Vermont, western Massachusetts and southeastern Connecticut. In central New York, those carbonates were unaffected, and from the late Silurian through the early Devonian, the Helderberg Group carbonates were deposited, shifting to clastics and back to the Onondaga Limestone by mid-Devonian. The onset of the Acadian Orogeny to the east in the late Silurian progressed westward, shedding clastics and eventually greatly deforming the Devonian carbonates along the Hudson River Valley. Other than Holocene calcareous tufa, no further carbonates were deposited on the region; clastic units of the Catskills migrated westward as the Alleghenian Orogeny ended the Paleozoic with the completion of the supercontinent of Pangea. Triassic rifting influenced southern New York and formed the Connecticut River valley, and Mesozoic intrusions developed farther east in New Hampshire and Maine. Erosion and peneplanation dominated the region from the Mesozoic through the Cenozoic until the onset of continental glaciation in the Pleistocene. Late Cenozoic doming brought the Adirondack Mountains upward, the region's oldest rocks becoming the youngest tectonic feature. Caves and karst as visible today may have initiated in the late Pliocene.

To understand the karst of the northeastern United States it is important to know the events that shaped the region, from the Precambrian and Paleozoic orogenic events, to the Quaternary glaciations. These events created the various physiographic provinces of New York and New England (Fig. 2.1), each with differing geology shaped by these events. Mountain building events in the Precambrian and Paleozoic have created the basement rocks of the region; developed basins in which sediments have been deposited, and deformed preexisting sedimentary rocks, in some cases metamorphosing them. Rifting in the region has also created

M.P. Cooper and J.E. Mylroie, *Glaciation and Speleogenesis*,
Cave and Karst Systems of the World, DOI 10.1007/978-3-319-16534-9_2,

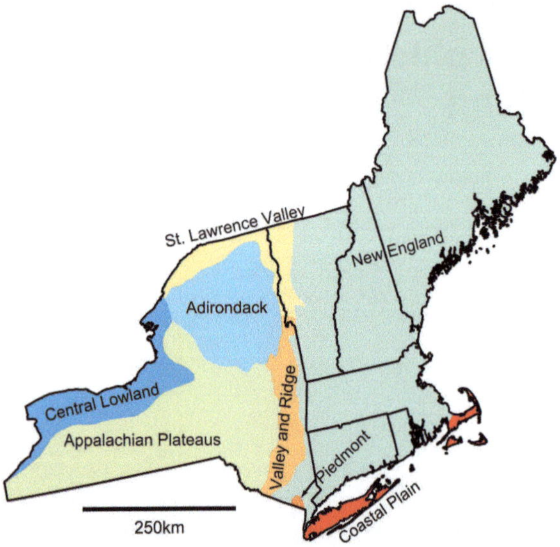

Fig. 2.1 The physiographic provinces of New York and New England follow their underlying geology. The Appalachian Plateaus province is characterized by flat lying, Siluro-Devonian strata in New York. The Valley and Ridge province contains deformed Siluro-Devonian strata, modified through multiple orogenic events. The lowland provinces that rim the Great Lakes, St. Lawrence River, and Lake Champlain contain relatively flat lying Cambrian, Ordovician, and Silurian strata, though some Precambrian exposures exist. The Adirondack province is composed of Precambrian metamorphic units. The New England province consists mainly of metamorphosed Paleozoic strata, as well as Mesozoic plutonic rocks. The Piedmont and Coastal Plain provinces do appear in the region, but are limited. These different provinces correspond to the differing karst regions explored here; the Helderberg Plateau region exists in the Appalachian Plateau, the Hudson Valley Fold-Thrust Belt region exists in the Valley and Ridge, and the northern New York Lowlands exist in the Central Lowland, St. Lawrence Valley, and Appalachian Plateaus. The Adirondacks and New England karst regions correspond to their respective provinces

basins during the breakup of Pangea, and preserves some sedimentary rocks from the Triassic Period. Igneous rocks have also formed from plutonism during mountain building and rifting. While igneous rocks do not form karst they do form pseudokarst and non-dissolution caves (Chap. 4). In the New England states with low karstic rock extent (Chap. 1, Fig. 1.1), non-dissolutional caves are the primary cave type.

Geomorphically New York and New England have been shaped by erosion of the previously high-relief mountain ranges, as well as the low relief plateaus of non-deformed areas. Erosion has included long-scale fluvial erosion, as well as the glacial erosion of the Quaternary.

Of importance for the discussion of the karst of northeastern United States are the creation of basins, the deposition of limestone and dolomite and their adjacent non-soluble units, the deformation of these units and their metamorphism, and the erosion of these units to their modern extent by fluvial and glacial processes.

2.1 Precambrian—Grenville Orogenic Events

The oldest rocks in New York and New England are the Precambrian Grenville-aged metamorphic and igneous rocks. These rocks make up the basement rock of much of the eastern United States as the Laurentian Craton, and crop out in New York in two areas: the Adirondack Mountains, and the Hudson Highlands (Isachsen et al. 2000). This suite of rocks is composed of metamorphosed sedimentary rocks that were deposited in the shallow pre-Grenville seas, metamorphosed anorthosites in the Adirondacks, and granites in the Hudson Highlands. The Grenville orogeny is associated with the formation of the supercontinent Rodinia (Fig. 2.2), and occurred in multiple events between 1.3 and 1.0 Ga (Tollo et al. 2004). Dates from

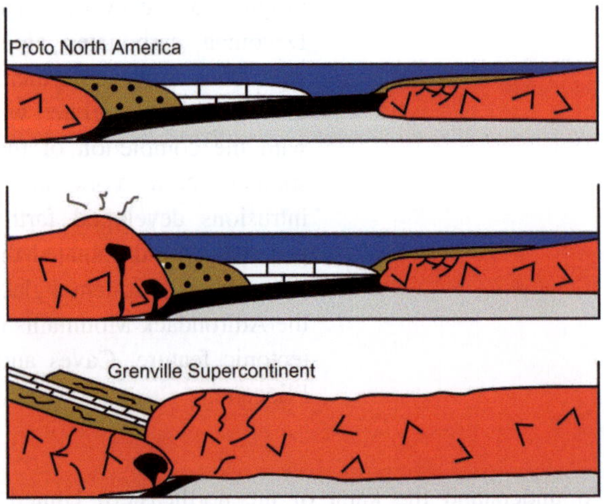

Fig. 2.2 Formation of the Grenville supercontinent during the Grenville Orogeny (1.3–1.1 Ga). Clastics and carbonates were deposited in the pre-Grenville ocean off of Proto North America. These units were metamorphosed and are the basement rock of much of eastern North America, outcropping in the Adirondack Mountains, as well as the Hudson Highlands

exposed rocks in New York associate them with the 1.1 Ga orogenic phase. While the rocks associated with the Grenville Orogeny are the oldest in the region, their exposure is recent in the Adirondack Mountains as dome uplift continues to this day, making them the youngest mountains in the region.

The variety of rocks within the Grenville metamorphics includes schist, gneiss, marble, and meta-anorthosite. Igneous rocks include a variety of granites that comprise the Hudson Highlands. The metamorphic rock units crop out in both southeastern New York, and in the Adirondacks, though these units have a larger extent in the Adirondacks. Caves and karst have formed in the marble distributed around the Adirondacks, and the marble units tend to be sandwiched between two schist units, analogous to the stripe-karst seen in Norway, though they do not produce the multi-level network mazes displayed in the Norwegian examples (Faulkner 2009; Chap. 9).

While not karst forming, the igneous and metamorphic rocks of the Grenville suite do produce non-dissolution caves. When over-steepened, these rock slopes can produce crevice caves on initial fracture, and talus piles and caves upon fracture failure (Chap. 4).

Additional exposed Precambrian rocks in New England include those of the Berkshire Massif, a series of metamorphic rocks that crop out in Connecticut, Massachusetts and Vermont. Other karst forming Precambrian units includes the Sherman Marble (Porter 2009) of Massachusetts and Vermont.

2.2 Paleozoic Mountain Building and Deposition

2.2.1 Cambro-Ordovician—Taconic Orogenic Events

After the Grenville orogeny, the subsequent period of the Rodinia supercontinent rifting occurred creating the Iapetus Ocean, and defined the beginning of the continent of Laurentia (North America and Greenland). This rifting created basins in which sediments were deposited and preserved. In some eastern parts of New York and in New England the Cambrian sedimentary rocks were later metamorphosed, including limestone into marble where karst has formed (Chap. 8). In the Lake Ontario—St. Lawrence—Lake Champlain lowlands, Mohawk Valley, and west of the Taconic Mountains along the Hudson River Cambro-Ordovician limestones crop out. Those Cambro-Ordovician units close to the Taconic Mountains have been deformed, but not metamorphosed, while those in northern New England, the northern New York lowlands and Mohawk Valley are minimally deformed (Isachsen et al. 2000).

During the Cambrian and Ordovician eastern Laurentia was covered in shallow, epicontinental seas in passive margin conditions. Sedimentation during this time can be broken into two sequences: the clastic Taconic Sequence, and the Carbonate or Shelf Sequence (Zen 1961; Isachsen et al. 2000). Initial sedimentation in the basins occurred from the erosion of Grenville-aged material generating quartz sands in both the carbonate/shelf sequence, and the Taconic sequence. The Taconic Sequence (Fig. 2.3) contains a variety of clastic formations with outcroppings in New York, Massachusetts, Connecticut, and Vermont. This sequence of clastics originates from the approaching

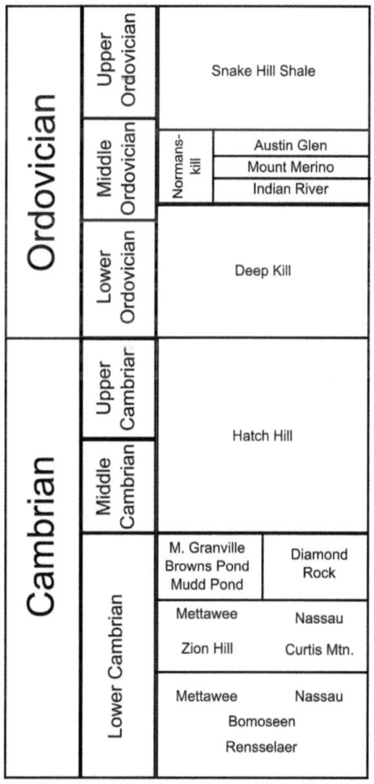

Fig. 2.3 Stratigraphic column of the Taconic sequence of clastics that eroded from the Taconic/Iapetus terrane. While there are no carbonate units in this clastic sequence, carbonates were deposited contemporaneously to the west

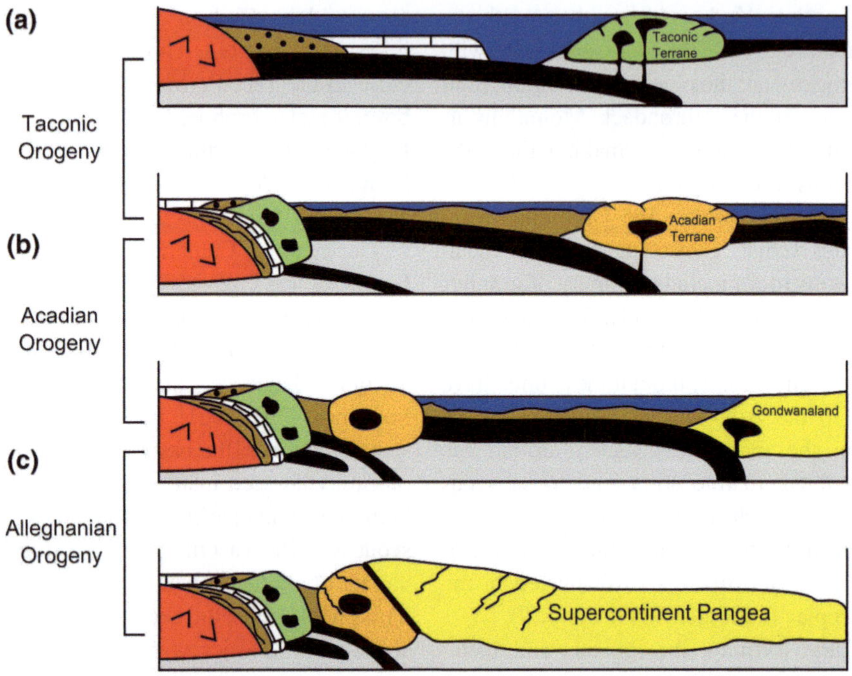

Fig. 2.4 The northeastern United States has been affected by several orogenic events. **a** The Taconic orogeny, involving the Grenville province and volcanic arc Taconic Terrane. **b** The Acadian orogeny, involving Laurentia, and the Acadian Terrane. The Acadian orogeny added parts of Connecticut, New Hampshire, and Maine, as well as the entirety of Rhode Island onto the eastern margin of Laurentia. **c** The Alleghenian orogeny occurred between Laurentia and Gondwanaland, closing the Iapetus Ocean and forming the supercontinent Pangea

volcanic Iapetus Terrane (Fig. 2.4a). As clastic input generally shuts down carbonate production, the Taconic Sequence is lacking in carbonate units, and therefore does not develop karst. Some of these continuous or equivalent units have differing stratigraphic names depending on the state, and some units have confusing multiple names due to their fragmentary outcrop nature resulting from deformation (Vollmer and Walker 2009).

Of greater interest for karst development, the Carbonate Sequence (Fig. 1.2) of the Cambrian and Ordovician generated several extensive limestone and dolomite units, some of which have been subsequently metamorphosed into marble. Paleogeographically, Laurentia was located along the equator at this time (Fig. 2.5), allowing the deposition of these carbonates in shallow epicontinental seas, and out to the paleoshoreline in modern day Vermont, Massachusetts, and Connecticut. In southeastern New York and western New England the earliest unit deposited in the Cambrian was the Stockbridge Group (called the Wappinger Group in New York). In the lowlands of northern New York and northwestern New England the Beekmantown Group was deposited contemporaneously

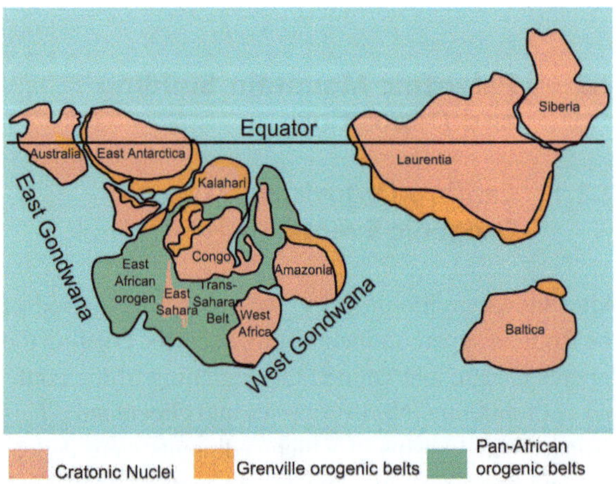

Fig. 2.5 Paleogeographic map for the Cambrian. During this time most of the continents were tied up near the South Pole. Laurentia's position is near the equator, allowing the formation of carbonates during this period, and subsequent time periods of the Paleozoic. (Redrawn from Technical Report USGS-TR-98-3)

with the Stockbridge Group. Karstic units within the Beekmantown Group include the Bascom Formation (Porter 2009) and the Little Falls Formation (Engel 2009). Both of these groups have karstic units (Plante 1990; Engel 2009), and the Stockbridge Group forms most of the karst in the Berkshires where it was metamorphosed into marble during the Taconic and subsequent orogenic events. Deposition of these units continued into the Lower Ordovician.

Through the Middle Ordovician, deposition of carbonates continued in the shallow seas of Laurentia. One of the most extensive carbonate units of the Middle Ordovician is the Chazy Group. Outcrops of the formations within the Chazy Group extend from the Champlain lowlands of northern New York, through Vermont, and into Newfoundland. This group represents some of the earliest fossilized reefs. The karst units within the Chazy Group such as the Valcour Limestone contain several caves in northeastern New York and western Vermont.

Through the Middle Ordovician the Taconic Orogeny began influencing eastern Laurentia causing greater clastic input along the paleo-shoreline. Throughout the beginnings of orogeny, however, some periods of tectonic quiescence existed, allowing deposition of carbonates of the Walloomsac Formation, an irregularly appearing karst forming formation ranging from modern day western Massachusetts to southeastern New York in the Stockbridge Valley (Plante 1990; Isachsen et al. 2000), and the karst forming Middlebury and Chipman formations in Vermont (Porter 2009; Scott 2013). Continued mountain building in this area during the Middle Ordovician formed the Taconic Mountains, and the Berkshire Mountains (termed Green Mountains in Vermont), and metamorphosed the clastics of the Taconic Sequence, and the carbonates of the Carbonate Sequence in eastern New York, western Massachusetts, and western Vermont (Fig. 1.3) producing a Barovian metamorphic sequence from the clastics (Vollmer and Walker 2009), and the Stockbridge Group and Walloomsac marbles.

Despite the mountain building towards the east, carbonate deposition continued in the northern lowlands and Mohawk Valley of New York, including the Black River Group and Trenton Group. These groups represent up to 140 m of carbonates (Isachsen et al. 2000), deposited in intertidal to deep environments. Both groups are karstic, and contain extensive maze caves in the Lake Ontario—St. Lawrence lowlands along the Black River, and Perch River, along with other smaller caves. These units are relatively non-deformed, but do contain some folds, and monoclines form important control on geomorphic development (Cushing et al. 1910), including karst development (Chap. 7).

The Taconic orogenic events ceased by 440 Ma (Isachsen et al. 2000), leaving the Iapetus Terrane accreted to Laurentia. This accreted terrane enlarged Laurentia and moved the paleo-shoreline to eastern Massachusetts and added substantial portions of New England including much of Connecticut and Vermont, as well as most of New Hampshire and Maine. At this time Greenland, northern Ireland, northwestern Scotland, and northern Scandinavia were also part of Laurentia (Torsvik et al. 1996). Sedimentation in basins during the Late Ordovician ceased in the western parts of the region as the Taconic and Berkshire Mountains were built and were eroding, but continued in western New York, depositing shales and sandstones such as Queenston Formation. Deposition during Late Ordovician and Early Silurian was mainly shut off ending the Sauk megasequence (Sloss 1963), due to eustatic sea-level lows associated with tectonics and Ordovician glaciation (Isachsen et al. 2000).

2.2.2 Siluro-Devonian—Acadian Orogenic Events

Sea level began to rise again in the Early Silurian, allowing marine sedimentation to occur and be preserved. Early Silurian units preserved in New York and New England outcrop in western New York, New Hampshire, Vermont, Massachusetts, and Maine. Early Silurian units in western New York along the Allegheny Plateau and Erie-Ontario Lowlands are those of the Medina and Clinton groups, with the Shawangunk Conglomerate in the east along the eastern edge of the Alleghany Plateau (Chap. 1, Fig. 1.2; Isachsen et al. 2000). The Medina Group and Shawangunk Conglomerate are both entirely clastic as erosion of the Taconic Mountains supplied vast amounts of clastic sediment, therefore shutting off carbonate production. The Shawangunk Conglomerate is over 300 m thick at its maximum, representing braided stream deposits of quartz sands and conglomerate. The Medina Group to the west contains units that are fluvial to offshore. The Clinton Group continues the trend of clastic rocks deposited in shallow and open shelf, though it does contain several carbonate units including the Irondequoit Limestone and the DeCew Dolostone that display limited karst development. In New England, Early Silurian deposition was also clastic, with some volcanics. These, like the New York strata are a mixture of braided stream to offshore facies (Thompson 1985), though unlike in New York the Acadian and Allegheny orogenies have metamorphosed the New England units.

In the Late Silurian clastic deposition continued in eastern New York, also represented by the thick Shawangunk Conglomerate. In the west carbonate production was reinitiated with the karst-forming Lockport Group (Engel 2009), containing limestone and dolomite. The Late Silurian also continued clastic deposition in New England. As the Late Silurian continued, sea-level transgression and low relief of the region changed clastic deposition to evaporative coastal plain and nearshore environments of the Salina Group in eastern and western New York and New England. The Salina Group in western New York contains evaporates including gypsum and rock salt, though no known evaporate karst exists within these rocks. Following this deposition, the Taconic Mountains were eroded to very low relief, shutting off clastic input and renewing carbonate deposition in eastern New York (Isachsen et al. 2000), and parts of Massachusetts and New Hampshire (Thompson 1985). Late Silurian carbonate units include the occasionally karstic Rondout Formation and early Helderberg Group in New York, and the Fitch Formation in Massachusetts and New Hampshire.

The beginning of the Devonian is marked by a transition to completely carbonate deposition in the form of the Helderberg Group (Fig. 2.6) in New York. Carbonate deposition also occurred in the Early Devonian of Massachusetts with the Gile Mountain and Waits River formations, and in Vermont with the Waits River Formation. The Early Devonian carbonate and clastic units of New England are, like most units in this region, metamorphosed. The Helderberg Group in New York contains several limestone formations above the Rondout Dolomite: the Manlius, the Coeymans, the Kalkberg, the New Scotland, the Becraft, the Alsen, the Port Ewen, and the Port Jervis. These units outcrop in the eastern border of the Allegheny Plateau, and along the northern border of the Allegheny Plateau, particularly in the Helderberg Plateau. These units represent several facies: the intertidal Manlius, the shallow shelf Coeymans, the deeper water, chert-rich Kalkberg, and the fossil rich, deeper water New Scotland (Isachsen et al. 2000). The Becraft, Alsen,

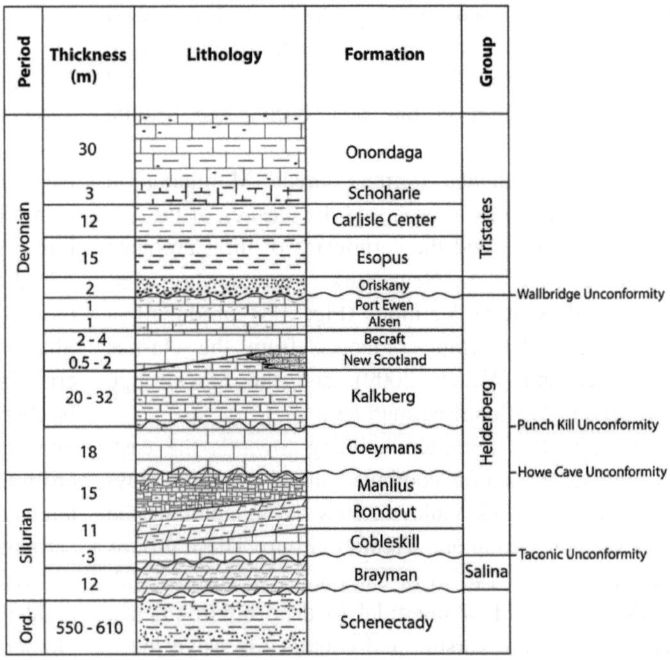

Fig. 2.6 Stratigraphic column of the Siluro-Devonian units outcropping in the Helderberg Plateau, and elsewhere in the Appalachian Plateaus. (From Weremeichik 2013)

and Port Ewen formations are a repeat of the Coeymans, Kalkberg, and New Scotland facies respectively. The Helderberg Group is continuous in the east representing the full transgressive, regressive, and the following additional transgressive stratigraphic sequence, but to the west in the Helderberg Plateau the New Scotland, Alsen and Port Ewen formations are not extant, or were never deposited (Mylroie 1977). The Port Jervis Limestone only exists in Port Jervis, New York and represents the New Scotland facies. The time-transgressive nature of the Helderberg Group westward means that the Silurian-Devonian boundary moves downward through the Manlius and Rondout formations as the marine transgression proceeded west (Ebert and Elliot 2003). There is some disagreement in the literature regarding whether or not the Rondout Formation is part of the Helderberg Group; the confusion is in part because the Lower Manlius is now recognized to be Silurian in the eastern part of New York, but transitions to Devonian proceeding west. Ebert and Elliot (2003) provide a review of the situation. Isachsen et al. (2000) treat the Rondout as separate from the Helderberg Group; from a cave and karst viewpoint, the Rondout behaves as part of the Helderberg Group.

The Helderberg Group, particularly the Manlius Formation contains the most well developed caves and karst in the northeastern US, as the units are extensive and thick. The Manlius and overlying Coeymans Formations are particularly well suited for karst development as the Manlius is a very pure, thin-bedded micrite and the Coeymans is a massive, soluble caprock (Mylroie 1977). Other units of the Helderberg Group do form karst and caves, but not as extensively as the Manlius and Coeymans Formations. High chert content in units of the Kalkberg facies especially do not form extensive karst, but provide insoluble material that can guide flow paths.

The Port Jervis Formation marks the end of the Tippecanoe megasequence, and thus begins a period of erosion and the Wallbridge Unconformity (Sloss 1963) due to eustatic sea level lows. This formation is one of the only extant units from this time period. Following the Port Jervis formation are those of the Tristates Group (Fig. 2.6), a mixture of clastics and carbonates. Units that were deposited and preserved as sea level began to rise again are terrestrial clastics, including the Oriskany Sandstone and Connelly Conglomerate, as well as the marine Glenerie Limestone (Kalkberg facies). The Oriskany Sandstone imposes an important control on karst formation, as this sandstone allows allogenic recharge to the limestones below, and can allow formation of sandstone-capped maze caves in the underlying Becraft Limestone (Cullen et al. 1979). These low sea-level clastics were preserved as sea level rose, burying them with marine sediments of the Esopus Formation (Isachsen et al. 2000). The Esopus Formation is a shale unit, containing bentonites formed from volcanic ash. The combination of volcanoclastics and erosional clastics in the Esopus Formation mark the first phase of the Acadian Orogeny. This orogeny involved Laurentia, the advancing Avalonia Terrane, and the attached continent of Baltica (Figs. 2.4b and 2.7; Torsvik et al. 1996). The Avalonian Terrane includes what are now parts of Connecticut and Maine, and the entirety of Rhode Island, southern Britain and Ireland, and western Europe. Baltica at this time included Fenno-Scandinavia, eastern Europe, and northwestern Russia.

Units above the Esopus Shale in the Tristates Group are a mixture of carbonates and clastics, with the Schoharie and Rickard Hill formations being a mixture of both. The Onondaga Formation is a karstic limestone unit with several members that are repeats of facies of the Helderberg Group: the Edgecliff Member (Coeymans), the Nedrow Member (New Scotland), the Moorehouse Member (Kalkberg), and the Seneca Member (New Scotland) (Isachsen et al. 2000). The Nedrow and Moorehouse members, like their respective Helderberg facies contain abundant chert. The Seneca Member contains the Tioga volcanic ash beds, again indicating the beginning of another phase of the Acadian Orogeny. The Onondaga Formation is another dominant karst former along the Allegheny Plateau, and extends farther west than the Manlius Formation, though it also forms caves in areas where the Manlius Formation exists such as the Helderberg Plateau.

After deposition of the Onondaga Formation, the Acadian Orogeny continued, causing mountain building in the east, and tacked on parts of Avalonia as Rhode Island, and parts of the eastern coast of Maine, Connecticut, and Massachusetts. The mountain building of the Acadian Orogeny caused the formation of the Acadian Mountains, the deepening of the Appalachian Basin, deformation and metamorphism of sedimentary rocks in New York and New England, and plutonism forming granites and other igneous

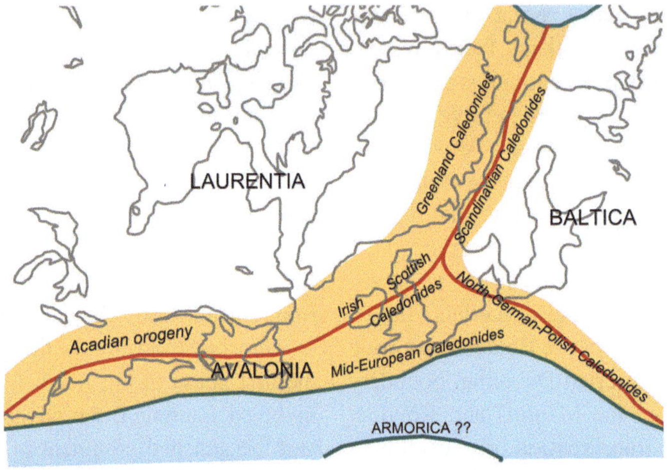

Fig. 2.7 The Caledonian/Acadian orogeny involving Laurentia, Avalonia, and Baltica. The Acadian orogeny added much of New England to the coast of Laurentia. The Caledonides in Ireland, Scotland, Greenland, and Scandinavia are a continuation of the Acadian mountain building event in those regions. This orogeny, combined with the Alleghenian has deformed and metamorphosed the sedimentary rocks of eastern North America, and the countries mentioned above. Map from Woudloper, using data from Torsvik et al. (1996). Image license is Creative Commons ShareAlike 1.0 [http://creativecommons.org/licenses/by-sa/1.0/]

rocks in Maine, New Hampshire, and eastern portions of the other New England states. Accompanying the Acadian Orogeny is the renewal of clastic deposition, a trend continued through the Late Devonian and into the later Paleozoic periods (Fig. 2.6). This deposition occurred in the Appalachian Basin in New York, as the rocks of New England eroded. Little to no preserved deposition exists in New England from the Middle Devonian to the end of the Paleozoic.

In the east deposition occurred in deltaic and fluvial environments in the form of the Catskill Delta, with deep water shales such as the Marcellus Shale to the west. As sea level dropped and sedimentation continued, facies migrated westward. In the eastern portions of the Allegheny Plateau hundreds of meters of this clastic material are preserved, tapering towards the west (Isachsen et al. 2000). Continued sea level fall, and subsequent rise, allowed the preservation of the large amount of terrestrial sediments. The return to sea level rise occurred after the deposition of the Hamilton Group, and continued deposition westward. Following this unconformity, sea level transgression shut off clastic input towards the west, allowing carbonate production briefly with the Tully Limestone. Some small caves have formed in the Tully Limestone in western New York. In the east terrestrial clastic deposition continued, with additional building of the Catskill Delta. Continued transgression and regression slowly moved facies westward, through the Late Devonian.

2.2.3 Mississippian, Pennsylvanian, and Permian—Allegheny Orogeny

Deposition continued to occur into the Mississippian, Pennsylvanian, and Permian periods in the Appalachian Basin. Little sedimentary rock from these periods remains, however, in the Allegheny Plateau of New York, as erosion during the Mesozoic and Cenozoic stripped these units away (Isachsen et al. 2000). Units from this time period in the Appalachian Basin do exist towards the south, including extensive carbonates from the Mississippian in the Virginias and to the south and west. Northwards clastic deposition continued in Pennsylvania and New York, with preservation to the current day in Pennsylvania. In New York, only minor Mississippian and Pennsylvanian units persist along the southern portion of the Allegheny Plateau in the state.

During the Mississippian another orogenic even began, the Allegheny Orogeny. This orogeny occurred between Euramerica (Laurentia, Avalonia, and Baltica) and Africa (at that point part of Gondwana), and began the formation of the supercontinent Pangaea

(Fig. 2.4c; Torsvik et al. 1996). Orogeny continued through the Pennsylvanian and into the Permian. In New England this orogeny increased the height of the Acadian Mountains, and continued deformation and metamorphism in the northeast. In New York deformation occurred in the southeast, affecting the eastern reaches of the Allegheny Plateau. This deformation includes carbonate units, and in this area deformation structures greatly influence cave development (Chap. 6). Westward in New York, however, little deformation occurred (Isachsen et al. 2000), though several structural effects exist, including those important to karst development (Kastning 1975; Mylroie 1977). Jointing and cleavages have developed in the carbonate units of the Allegheny Plateau (Marshak and Engelder 1985; Isachsen et al. 2000). Cave development in this region is highly influenced by jointing (Chap. 5), as joints guide flow paths of initial cave enlargement (Palmer 1991).

The Late Permian in the northeastern US lacks a sedimentary record, as glaciation lowered sea level globally (Knoll et al. 1997), and the relief along the mountains formed by the Allegheny Orogeny lowered.

2.3 Mesozoic Rifting and Deposition

2.3.1 Triassic and Jurassic—Rifting

During the Triassic rifting began around 220 Ma (Isachsen et al. 2000), breaking apart Laurentia from Africa, forming the Atlantic Ocean. During the period of rifting failed rift basins formed (Fig. 2.8), as faulting allowed blocks to move downward. Along with this rifting, diabases and basalts were formed and preserved in these basins, such as the Palisades Sill diabase along the southern Hudson River, along with deposition of terrestrial clastics including redbeds of sandstone, conglomerate, siltstone, and shale. Basins of this time include the Newark Basin, the largest of the failed rift basins, as well as the Hartford, New York Bight, and other basins. Deposition continued through the Jurassic, including flood basalts (McHone 1996) in Connecticut, Maine, Massachusetts, Rhode Island, and New York. Elsewhere in New York and New England, high relief conditions continued.

After the failed rifts of the Triassic, the Atlantic Ocean began to fully open in the Early Jurassic, beginning deposition off of the northeastern coast, continuing into the Cretaceous. Erosion continued during this time, forming a hypothesized peneplain in New York during the Middle Jurassic (Isachsen et al. 2000). Jurassic granites also formed in New Hampshire as a result of the New England hotspot (Duncan 1984).

2.3.2 Cretaceous—Plate Movement, Erosion, and Deposition

During the Cretaceous the Atlantic Ocean continued enlarging, and active margin conditions transitioned to passive margin conditions. Erosion of high relief areas of New York and New England continued, depositing sediments in the Atlantic Ocean along the eastern coast in the Coastal Plain region. Most deposition in this time period occurred in the warm seas south along the east coast of North America, though some Cretaceous deposits exist along the New York and New

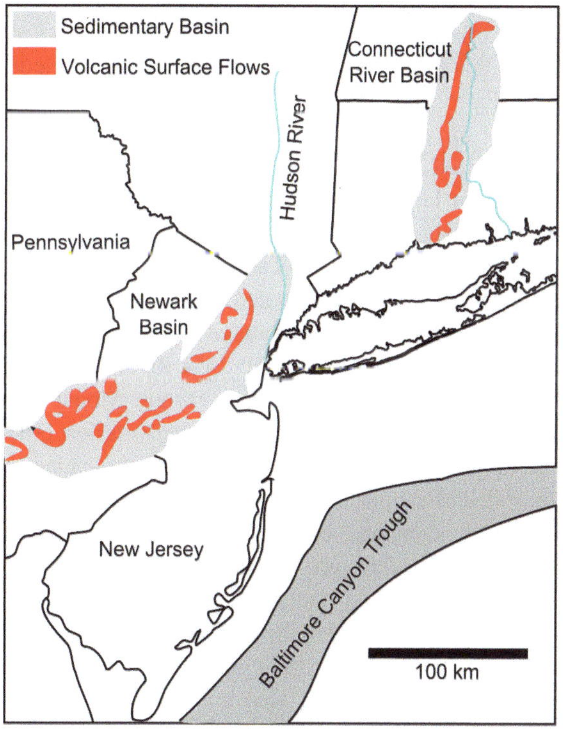

Fig. 2.8 During Triassic rifting basins were formed along the eastern margin of North America. These basins contain clastic sedimentary deposits, mainly red sandstones, as well as volcanic lava flows and dikes. (Modified from USGS [http://3dparks.wr.usgs.gov/nyc/mesozoic/mesozoicbasins.htm], accessed 2015-01-20)

England coast, including the original sedimentation underlying Long Island. Erosion continued through the Cretaceous as the region was uplifted, beginning dissection of the hypothesized Middle Jurassic peneplain (Isachsen et al. 2000).

By the end of the Cretaceous North America was mostly the same shape as it is now, as rifting continued in other areas and Pangaea broke apart. During this movement North America continued to move over the New England hotspot, forming the White Mountains due to plutonism (Duncan 1984), and the New England (also known as the Kelvin) Seamount chain to the east in the Atlantic Ocean.

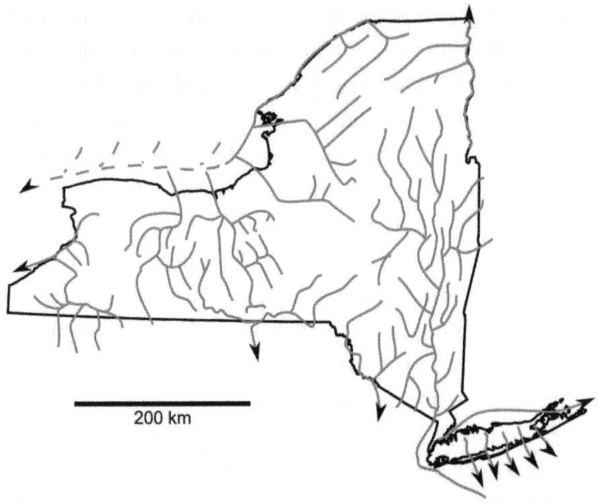

Fig. 2.9 An interpretation of the pre-Quaternary drainage of New York State. This loose interpretation is based on remaining, relict stream valleys. The Hudson River maintains much of its pre-glacial course. The Finger Lakes and Lake Champlain are interpreted to be former river channels. (Adapted from Isachsen et al. 2000)

2.4 Cenozoic Geomorphic Development

2.4.1 Paleogene and Neogene—Erosion

Few deposits exist from the Paleogene and Neogene in New York and New England, as erosion continued during this time. Those deposits that do exist include early Paleogene lignite deposits in Vermont. The erosion of this time continued to downcut the hypothesized Jurassic peneplain, establishing much of the current large drainage patterns (Fig. 2.9). These drainage patterns reflect the underlying geology established during the Precambrian, Paleozoic, and Mesozoic. Where structural deformation exists gradients are high, and channel drainage moved into weaker units, as exemplified by the Hudson River following the western edge of the weathered Taconic Mountains. In the Allegheny Plateau drainage patterns reflect little structural control and thus are mostly stratigraphically controlled. This plateau was dissected, forming sedimentary “mountains” by removing material, causing isostatic rebound, in turn causing more erosion.

The erosion of the Paleogene and Neogene drove additional uplift, bringing the current stratigraphy now seen in the northeastern US to the surface. Most of the current bedrock exposure seen today was carved out due to this erosion, except for possibly the continually uplifting Adirondack dome (Isachsen et al. 2000).

The Late Neogene is marked by a decrease in global temperatures, as they dropped from the Cretaceous and through the Paleogene. Late Neogene cooling, plate arrangement, and ocean currents were set on course for the “just right” conditions to begin an ice age, similar to those seen before in the Proterozoic, Ordovician, and Permian.

2.4.2 Quaternary—Ice Age

The Quaternary Period is a time of ice ages following the Neogene (Chap. 1). In New York and New England there are multiple glacial, deglacial, and interglacial cycles, the most recent being the Wisconsinan glaciation, an advance of the Laurentide ice sheet covering Canada, almost all of New York and all of New England, northern New Jersey and northern Pennsylvania, and other northern US states (Fig. 2.10). The Quaternary is broken into the Pleistocene, and the current interglacial, the Holocene.

2.4.2.1 Pleistocene—Glacials and Interglacials

The Pleistocene contains several glacial and interglacial time periods. In eastern portions of North America the most recent are named, from youngest to oldest: the Wisconsinan glacial, the Sangamonian interglacial (MIS 5e, Chap. 1), the Illinoian glacial, and the Pre-Illinoian glacials and interglacials (Muller and Calkin 1993). The geomorphologies of Pre-Illinoian cycles have been overridden by the subsequent Illinoian and Wisconsinan glaciations, and are pieced

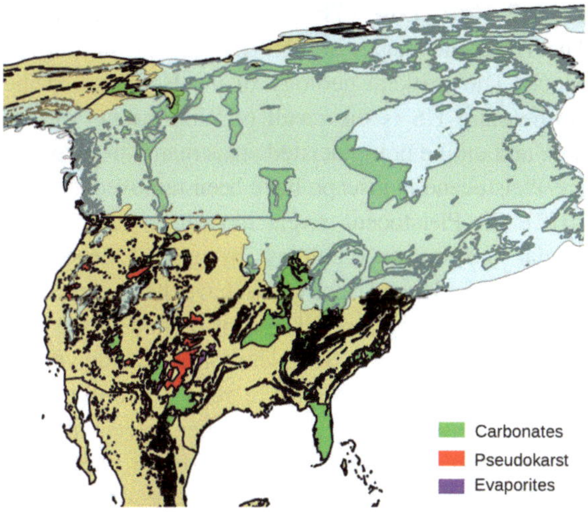

Fig. 2.10 Map of ice extent during the last glacial maximum (LGM) in North America, combined with potentially karstic and pseudokarstic rock exposure. All karst regions in Canada have been glaciated, either by continental glaciation from the Laurentide ice-sheet, or by alpine glaciation in the Canadian Rockies. The entirety of New York and New England was glaciated during the previous (Wisconsinan) glaciation, as well as further south in New Jersey and Pennsylvania, though karst regions of these states were not overridden by ice. Other northern states in the US were continentally glaciated, including those containing karstic rocks; continentally glaciated karst exists in states such as Michigan. Alpine glaciation also affected karst development in the Rocky Mountains, in states such as Colorado, and in areas such as the Tetons

together from marine isotope data. Both the Illinoian and Wisconsinan glaciations can be divided into further substages, with advance and retreat of various ice lobes. Glaciation here will be treated as a whole for pre-Wisconsinan as the Wisconsinan glaciation in New York and New England has reset prior surficial geology almost entirely. The Illinoian glaciation is known to have taken place in New York and New England as landforms associated with this glaciation exist southward in New Jersey, beyond the Wisconsinan glacial limit.

Large-scale glacial landforms in the northeastern United States were likely shaped over multiple glacial advances, such as the Great Lakes, Finger Lakes, and Long Island. The multiple glacial advances slowly eroded more and more material away, causing uplift bringing older rocks to the surface, even to the basement Grenville rocks in the Adirondacks. This erosion can remove shallow caves and cave passage, though the process can also bring more soluble rock to the surface. Smaller scale glacial landforms of previous glacial advances are reset upon subsequent glaciation, and as such the current erosional and depositional features represent the last glacial advance (Muller 1977), unless protected outside of the last glacial limit. The removal of glacial landforms makes it nearly impossible to precisely date and locate ice-limits of previous glaciations, though large-scale advance and retreat can still be dated using ice-core data, as well as deep-sea sediment core data. Speleothem data can also offer some timing controls.

While surficial deposits of previous glaciations were removed by subsequent glaciation, glacial deposits can be preserved in caves (Palmer 1975; Mylroie 1984; Weremeichik and Mylroie 2014). Some of these deposits can be dated to pre-Wisconsinan time (Dumont 1995; Lauritzen and Mylroie 2000; Perzan et al. 2014). As caves are preservational environments it may be possible to date and locate previous ice-limits using preserved material. Dumont (1995) and Lauritzen and Mylroie (2000) calculated age-dates of speleothems using U/Th age dating. Using this method it may be possible to date ice-advance and retreat as advance shuts off speleothem growth, and growth is renewed upon retreat (Lauritzen and Mylroie 2000). These methods are limited however, as they only apply to karst areas, and Lauritzen and Mylroie (2000) note that their data were not abundant enough to precisely map ice-advance and retreat, but it may still be a possibility if more data were collected from a larger range of sites. They were able to recognize a mid-Wisconsinan interstadial (MIS 3) as speleothem growth resumed for a brief time in the Hudson River valley, starting before, and ending later, than an even shorter speleothem growth pulse to the north in the Helderberg Plateau, during that interstadial.

The appearance of landscapes during previous interglacials likely resembled the current post-glacial (used here to mean Holocene) time period, with similar on-going processes.

The multiple readvances and retreats, and ultimate deglaciation of the Wisconsinan glaciation can be accurately mapped, and timed. In New York and New England multiple ice-lobes existed from the Laurentide ice sheet, including the Ontario, Hudson, Champlain, Connecticut Valley, Buzzards Bay, and Cape Cod lobes (Ridge et al. 1991; Dyke et al. 2002). These individual lobes also contain multiple sub-lobes. Ice flow direction of these lobes can be accurately mapped by drumlin orientation (e.g. Dineen 1986), and ice-advances and retreats can be mapped by end-moraine positions. Pollen stratigraphy and paleomagnetics can be used to date end-moraines, and associated times of advances and readvances (e.g. Ridge 2004). The last glacial maximum of the Wisconsinan glaciation occurred 28–24 ka, and glaciers had completely retreated from New York and New England by 12 ka (Ridge 2004). Ridge (2004) mapped deglaciation times for New England, giving accurate timings for ice-retreat (Fig. 2.11).

Glacial landforms that persist into the Holocene in New York and New England are associated with the Wisconsinan glaciation and deglaciation. These landforms are numerous, and exist at a variety of scales. The most important glacial features in the discussion of karst development are glacially downcut valleys and infilled valleys, which both change base level, but in different directions (Palmer 1976), and deranged drainage guided by erosional and depositional landforms, as these affect the recharge of cave systems, both of adjusted pre-glacial (used here to mean pre-Wisconsinan) caves (Mylroie and Carew 1987), and post-glacial caves (Cooper and Mylroie 2014).

2.4.2.2 Holocene—The Current Interglacial

The Holocene began at 11.7 ka, and is the current epoch. The surficial geology of the Holocene in the northeastern US is the result of Wisconsinan glaciation, and unlike non-glaciated areas many forms of the pre-Pleistocene landscape have been removed (Muller 1977). Pre-Pleistocene rivers have been rerouted by glacial erosion or damming by glacial deposition (Fig. 2.12). Several landforms have been preserved however, allowing a rough reconstruction of the pre-Pleistocene landscape. Such landforms include valleys that have been preserved as a result of infilling by sediment (e.g. Palmer 1976). Larger rivers, like the Hudson River and the St. Lawrence River likely have similar paths to their pre-glacial state, though the drainage basin of the St. Lawrence has been adjusted by glaciation (e.g. Cushing et al. 1910). Smaller streams may also maintain a general course, such as Onesquethaw Creek (e.g. Dineen 1987).

The Holocene time in the northeastern US displays a return to erosion and deposition by fluvial processes, rather than by glaciation, as such is the style in many non-glaciated terrains. On the small scale, routing of water is guided through deranged drainage by glacial landforms, and this water downcuts the landscape.

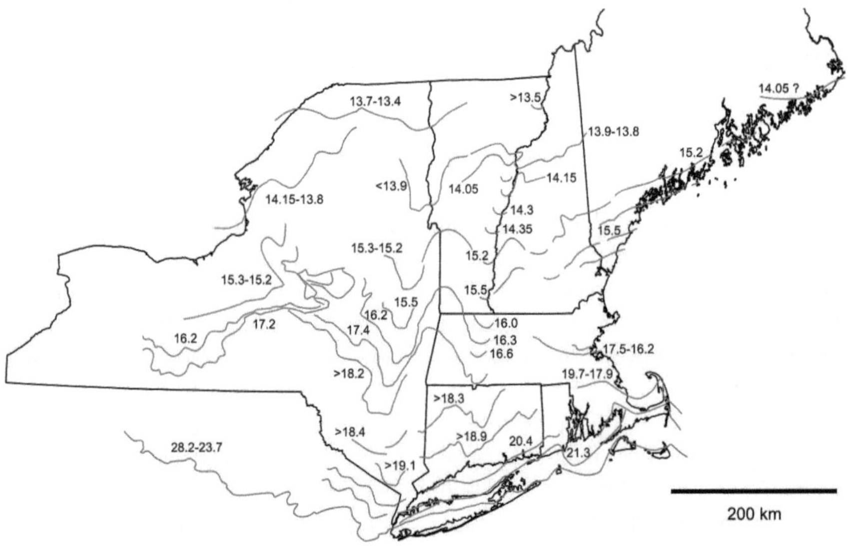

Fig. 2.11 Map of deglaciation of the northeastern United States. Gray isolines indicate time of withdrawal of ice-lobes indicated by end moraines. These dates are adjusted from calibrated ^{14}C dates of pollen located within moraines, and are in ka BP. The terminal moraine indicates the LGM in this region, and forms part of Long Island, New York. Ice-withdrawal from the region was complete by 13 ka BP. (Redrawn with data from Ridge 2004)

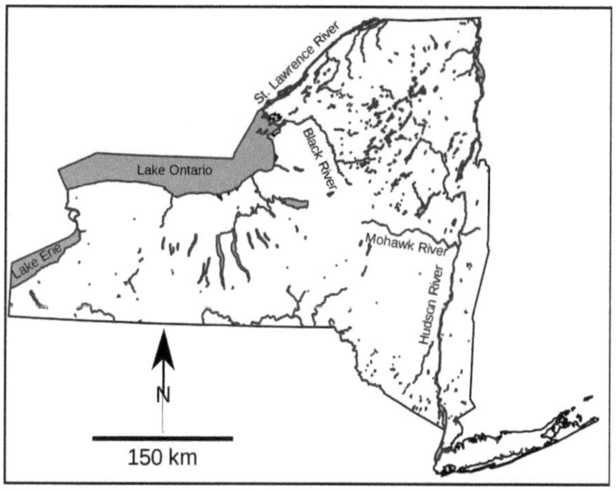

Fig. 2.12 Map of the modern drainage of New York State, showing modification by repeated glaciation. An example of this is the Finger Lakes, which were once river valleys (Fig. 2.9) but have been adjusted by glaciation. Several rivers such as the St. Lawrence and Hudson rivers have largely maintained their course, though with some adjustment

Sediment and organic material are also deposited in the many lakes of the post-glacial landscape, including within swamps caused by inefficient drainage of glacially carved depressions. Quaternary deposits in New York and New England are Holocene in age, or are from the Wisconsinan glaciation (Muller 1977). In some cases till deposits have been well compacted, though their surficial nature leaves them prone to removal by subsequent glaciation.

While glaciers are not in-place over the northeastern US during the Holocene, the effects of the glaciers still impacted the early Holocene. When the ice-sheets were in place, the crust was isostatically depressed. Removal of the ice-sheets, as well as the erosion provided by advancing ice, removed mass, and therefore caused isostatic rebound. Rebound has important impacts in post-glacial geomorphology, as it expands joints mechanically (Harland 1957), uplifts the terrain causing rapid fluvial downcutting, and can cause neotectonic earthquakes. Rebound, combined with glacial forms such as over-steepened slopes also influence post-glacial geomorphology as these can cause slope failure, and create features such as talus piles, common in the northeastern US in mountains, and along escarpments. Lake Champlain was converted from a marine inlet to a fresh-water lake by isostatic rebound.

While during interglacial times New York and New England are shaped by the same geomorphic processes as non-glaciated areas, the surficial geology, and thus karst development, is shaped by glaciation. The current surficial and shallow subsurficial forms in the northeastern US are transient in nature, and are limited in time to the relatively short interglacials.

References

Cooper MP, Mylroie JE (2014) Post-glacial speleogenesis: verification of a hypothetical model, and the origins of maze caves in glaciated terrains. Cave and Karst Sci 41(2):84–95
Cullen J, Mylroie JE, Palmer AN (1979) Karst hydrogeology and geomorphology of eastern New York. National Speleological Society, Huntsville
Cushing HP, Fairchild HL, Ruedemann R, Smyth CH (1910) Geology of the Thousand Islands region. New York State Museum Bulletin, no 145
Dineen RJ (1986) Deglaciation of the Hudson Valley between Hyde Park and Albany, New York. In: Cadwell DH (ed) The Wisconsinan stage of the first geological district, Eastern New York, pp 89–108
Dineen RJ (1987) Preglacial and postglacial drainage of the central Hudson Valley. In: Field trip guidebook, New York State Geological Association, pp B-1–B-30
Dumont KA (1995) Karst hydrology and geomorphology of Barrack Zourie cave system, Schoharie County. Bulletin V of the New York Cave Survey, New York
Duncan RA (1984) Age progressive volcanism in the New England Seamounts and the opening of the central Atlantic Ocean. J Geophys Res 89(B12):9980–9990
Dyke AS, Andrews JT, Clark PU et al (2002) The Laurentide and Innuitian ice sheets during the Last Glacial maximum. Quat Sci Rev 21:9–31
Ebert JR, Elliot NA (2003) Paleontology and Stratigraphy of the middle Helderberg Group at Rickard Hill Road. In Johnson EL (ed) New York State Geological Association Field Trip Guidebook 75th Annual Meeting. Schoharie, New York, pp. 197–204
Engel TD (2009) New York State. In: Palmer AN, Palmer MV (eds) Caves and karst of the USA. National Speleological Society, Huntsville, pp 27–32
Faulkner T (2009) Speleogenesis in New England marble caves. In: Proceedings of the Fifteenth International Speleological Congress, Kerrville, vol 2, pp 855–862
Harland W (1957) Exfoliation joints and ice action. J Glaciol 3 (21):8–10
Isachsen YW, Landing E, Lauber JM et al (2000) Geology of New York: a simplified account, 2nd edn. New York State Museum, Albany
Kastning EH (1975) Cavern development in the Helderberg Plateau, East-Central New York. Bulletin I of the New York Cave Survey, New York

Knoll AH, Bambach RK, Canfield DE, Grotzinger JP (1997) Comparative Earth history and Late Permian mass extinction. Science 273:452–457
Lauritzen S-E, Mylroie JE (2000) Results of a speleothem U/Th dating reconnaissance from the Helderberg Plateau. New York. J Cave Karst Stud 62(1):20–26
Marshak S, Engelder T (1985) Development of cleavage in limestones of a fold-thrust belt in eastern New York. J Struct Geol 7:345–360
McHone JG (1996) Broad-terrane Jurassic flood basalts across northeastern North America. Geology 24(4):319–322
Muller EH (1977) Late glacial and early postglacial environments in western New York. Ann N Y Acad Sci 288:223–233
Muller EH, Calkin PE (1993) Timing of Pleistocene glacial events in New York State. Can J Earth Sci 30:1829–1845
Mylroie JE (1977) Speleogenesis and karst geomorphology of the Helderberg Plateau, Schoharie County. Bulletin II of the New York Cave Survey, New York
Mylroie JE (1984) Pleistocene climatic variation and cave development. Nor Geogr Tidsskr 38:151–156
Mylroie JE, Carew J (1987) Field evidence for the minimum time for speleogenesis. Nat Speleol Soc Bull 49(2):67–72
Palmer AN (1975) The origin of maze caves. Natl Speleol Soc Bull 34(3):56–76
Palmer AN (1991) Origin and morphology of limestone caves. Geol Soc Am Bull 103:1–21
Palmer MV (1976) Ground-water flow patterns in limestone solution conduits. Thesis, State University of New York at Oneonta
Perzan Z, Munroe JS, Amidon WH (2014) Investigation of last interglacial sediment in Weybridge Cave, Vermont, USA. In: Geological society of America, abstracts with programs, vol 46, no 6, p 349
Plante AR (1990) Berkshire caves and karsts, berkshire geology and karst landforms, part-I: history, stratigraphy, and structure. Northeastern Caver 21(2):48–56
Porter C (2009) New England. In: Palmer AN, Palmer MV (eds) Caves and karst of the USA. National Speleological Society, Huntsville, pp 24–26
Ridge JC (2004) The Quaternary glaciation of Western New England with correlations to surrounding areas. In: Ehlers J, Gibbard PL (eds) Quaternary glaciations—extent and chronology—Part II: North America. Developments in quaternary sciences, vol 2. Elsevier, Amsterdam, pp 169–199
Ridge JC, Franzi DA, Muller EH (1991) Late Wisconsinan, pre-Valley Heads glaciation in western Mohawk Valley, central New York, and its regional implications. Geol Soc Am Bull 103(8):1032–1048
Scott M (2013) Stratigraphic evidence for tectonically driven subsidence during the deposition of the Middlebury formation (Middle Ordovician). Thesis, University of Vermont, Western Vermont
Sloss LL (1963) Sequences in the cratonic interior of North America. Geo Soc Am Bull 74(2):93–114
Thompson PJ (1985) Stratigraphy, structure, and metamorphism in the Monadnock quadrangle. Department of Geology and Geography, University of Massachusetts, Amherst
Tollo RP, Corriveau J, McLelland J et al (2004) Proterozoic tectonic evolution of the Grenville orogen in North America. Geological Society of America Memoir 197, Denver
Torsvik TH, Smethurst MA, Meert JG et al (1996) Continental break-up and collision in the Neoproterozoic and Palaeozoic—a tale of Baltica and Laurentia. Earth Sci Rev 40:229–258
Vollmer FW, Walker J (2009) The classic Barrovian metamorphic sequence of Dutchess County and it's structural and stratigraphic context in the Taconic Orogeny. In: Vollmer FW (ed) 81st annual meeting, field trip guidebook, New York State Geological Association, pp 11.1–11.20
Weremeichik JM (2013) Paleo environmental reconstruction by identification of glacial cave deposits, Helderberg Plateau, Schoharie County, New York. Thesis, Mississippi State University
Weremeichik JM, Mylroie JE (2014) Glacial Lake Schoharie: an investigative study of glaciolacustrine lithofacies in caves, Helderberg Plateau, Central New York. J Cave Karst Stud 76 (2):127–138
Zen E (1961) Stratigraphy and structure at the north end of the Taconic Range in west-central Vermont. Geo Soc Am Bull 72:293–338

Studies on Glaciated Karst in New York and New England

3

Abstract

The modern scientific study of glaciation, speleogenesis and karst development began in New York in the 1960s and 1970s, with major contributions from Vic Baker, Steve Egemeier, Ernst Kastning, John Mylroie, and Arthur and Margaret Palmer, all working in the Helderberg Plateau of central New York. They demonstrated that large, pre-glacial cave systems existed, and these caves had been modified by glaciation, primarily as a result of base-level changes, backflooding, and sediment occlusion. Their work and subsequent efforts in the 1980s and 1990s hypothesized post-glacial cave development as an outcome of deranged surficial drainage. The 1990s and 2000s saw new cave discoveries as a result of application of the previous glaciated karst models, and the refinement of karst drainage basins by dye tracing, which also lead to new cave discoveries. The application of U/Th dating to caves in the Helderberg Plateau demonstrated conclusively that the major cave systems had survived multiple glaciations in the Pleistocene. Comparison between marble caves in Norway and those in the northeastern US demonstrated much commonality of form and speleogenesis. Recent work has demonstrated glaciolacustrine deposits in caves, and that shallow maze caves of this region are post-glacial in origin. Research moved beyond the Helderberg Plateau region to the glaciated marbles of western New England, where post-glacial caves appear to dominate. The major impact of glaciation is now viewed as significant base level changes affecting flow routes, joint activation by glacial unloading, and sediment deposition in caves that creates additional backflooding.

While studies were going on in the classic continental glaciated karst areas of Britain, Canada and Norway in the 1960s and 1970s (Chap. 1), work was also ongoing in the northeastern United States, particularly the Helderberg Plateau in central New York (Chap. 5, Fig. 5.1). These investigations, as well as more recent investigations, have followed evolutionary trends in thoughts on glaciated karst. Additionally, these studies gave an updated view on glaciated karst in the United States, changing the perception of this area as a cave and karst region (Mylroie and Mylroie 2004). Recreational caving became very popular after the Second World War, with cavers collecting scientific data accompanying exploration (e.g. Fig. 3.1 and references therein). These efforts were published as products of the Northeastern Regional Organization (NRO)

M.P. Cooper and J.E. Mylroie, *Glaciation and Speleogenesis*,
Cave and Karst Systems of the World, DOI 10.1007/978-3-319-16534-9_3,

of the National Speleological Society, as true scientific studies (e.g. Gurnee et al. 1961) and regional compilations began appearing, such as *Caves of Albany County, NY* (Schweiker et al. 1960) and *Caves of Massachusetts* (Hauer 1969).

From these early studies work has followed both in the realms of caving, and in the science. Cavers in the Helderberg Plateau used the information gathered in these scientific studies as a predictor of cave locations, enabling them to find large, pre-glacial cave systems (Chap. 5; Dumont 1995). Some of these cave systems have required large digging efforts to remove glacial sediment occluding them, similar to the other classic continentally glaciated karst regions of Great Britain. The work of these cavers is chronicled in the regional caving newsletter, *The Northeastern Caver*.

Scientific studies in the Helderberg Plateau continued through the 1980s, 1990s, and into recent times. The studies of these decades updated the work of the 1970s, added supporting data, revised interpretations, confirmed hypotheses and continued following global trends on thought processes regarding glaciated karst, including comparisons to other glaciated karst regions (Faulkner 2009). New work also brought new techniques; such as U/Th age dating of speleothems into the fold (Lauritzen and Mylroie 2000), confirming fieldwork interpretations of cave timing.

The work performed in the Helderberg Plateau, and elsewhere in New York focused on several facets of glaciated karst, including timing of karstification, base level controls, flow path controls, and the effects of deposition, erosion, and sedimentation on caves and karst; and the respective interrelation of these facets. Here we will explore these studies time-wise, as well as by each of these facets and what the karst of the northeastern United States revealed about them.

Work on glaciated karst in New York and New England continues, with opportunities in not just the Helderberg Plateau, but also the rest of the northeastern US. The areas other than the Helderberg Plateau have a dearth of studies, though recent ones (Faulkner 2009; Quick 2010; Cooper and Mylroie 2014; Perzan et al. 2014) have extended into other regions in New York, and into New England.

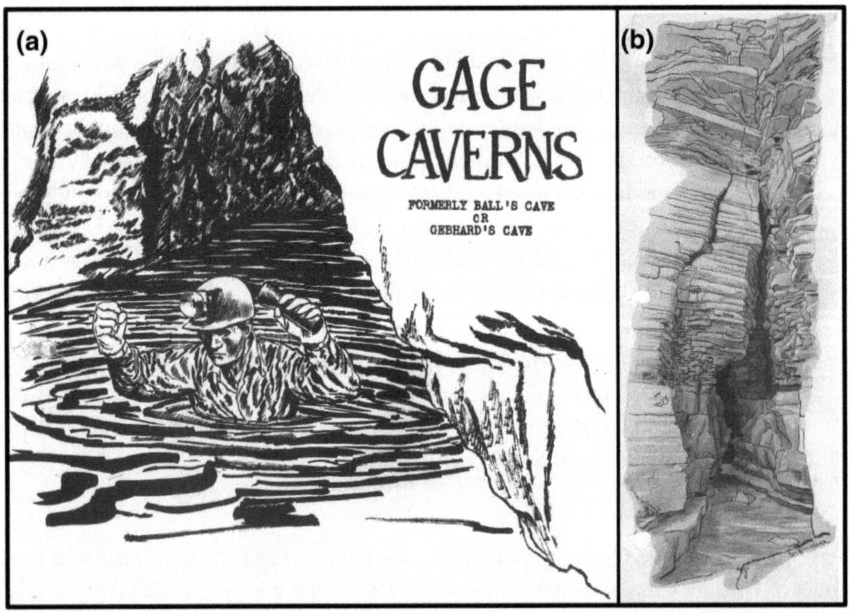

Fig. 3.1 Drawings from post-World War II caving publications in the northeastern US. **a** Lake Room, Gage Caverns, Schoharie County, New York (see Chap. 5, Fig. 5.6) (from Gurnee et al. 1958). **b** Entrance to Spider Cave, Schoharie County, New York (see Chap. 5, Fig. 5.6) (from Anonymous 1955). Both images were on the cover of the published documents. The drawings are by John Schoenherr, who later became a well-known illustrator, recognized for the classic original illustrations for the serialization of Frank Herbert's famous science fiction novel, *The Prophet of Dune* (published in book form as simply *Dune*), in Analog magazine in 1965 in five parts, January through May issues

3.1 Timeline and Impacts of Glaciated Karst Studies

The glaciated karst studies in the northeastern United States have occurred in several waves, interspersed with studies mainly limited to conference proceedings, field trip guidebooks, and publications in *The Northeastern Caver*. The published (and some unpublished) scientific studies are the result of several theses and a dissertation, as well as several papers by established academics at the time of their studies.

3.1.1 Late 1960s-Early 1980s

The first of these waves occurred from the late-1960s to the early-1980s, coincident with those in other continentally glaciated karst areas (Chap. 1). Studies prior to those pertaining particularly to glaciation are descriptive of the hydrology of the limestones of New York (e.g. Berdan 1948, 1950), as well as rough establishment of the origin of eastern New York caves (Egemeier 1969). The workers in the region during the 1970s and 1980s include those who have made great impacts on karst (both glaciated and non-glaciated), as well as on other fields. These workers include Victor Baker, Ernst Kastning, John Mylroie, Art Palmer, and Margaret Palmer. The studies these workers performed focused on the timing of cave development in glaciated areas, the effects of glaciation on caves, and reconstructing flow routes through cave systems. The work performed during this time period was mainly located in the Helderberg Plateau of Albany and Schoharie counties in east central New York (Chap. 5).

The works by Palmer (1972, 1975) are broad impact studies relating to maze caves of multiple varieties. Palmer's (1972) work on Onesquethaw Cave was one of the earliest studies in this region, in which he hypothesized a conceptual model to determine post-glacial origins of caves that was later examined by Mylroie and Carew (1987), and Cooper and Mylroie (2014). Palmer (1975) chiefly discussed the conditions under which maze caves form in all settings, but also includes influences of glacial sediment causing damming, the raising of hydraulic head, and the establishment of floodwater maze passages in pre-glacial cave systems. Earlier work by Palmer (1962) also includes the description, mapping, and speleogenetic history of Knox Cave (Chap. 5, Fig. 5.1).

The studies by Victor Baker (1973, 1976) are also some of the earliest, with fieldwork taking place in the late 1960s. Work by Baker (1973, 1976) includes dye tracing, analysis of well data (both hydraulic and water quality), and descriptions of caves and cave sediments. He also used the morphologic cave descriptions as well as cave sediments to describe paleohydrology and used these data to describe effects of glaciation on hydrology, including the claim of pre-glacial origins for the caves.

Ernst Kastning (1975) contributed additional dye tracing, as well as to the description and mapping of caves. Kastning (1975) increased the surveys of many known caves including McFails Cave (Chap. 5, Fig. 5.2), and Skull Cave (Chap. 5, Fig. 5.9). Kastning also mapped surficial features to narrow down their origin between karstic and glacial, demonstrating draining of these features by pre-glacial caves rather than by the establishment of new, post-glacial caves.

Palmer (1976) continued the trend of mapping caves, and also used water well log analysis. Also thrown into the mix of techniques by Palmer (1976) were gravimetric and electroresistivity geophysical studies. The mapping performed by Palmer (1976) included extremely detailed leveling surveys, with precise measurements of elevation, thus relating cave levels to previous base levels. Using the gravimetric and electroresistivity studies, as well as well data analysis showed that these base levels aligned with pre-glacial valleys infilled with glacial till (Chap. 5, Figs. 5.7 and 5.14).

Mylroie (1977, 1984) added the finishing touches on the early, glaciated karst studies in the Helderberg Plateau, adding more field evidence for pre-glacial cave origins and the survival of cave systems over multiple glacial cycles. Mylroie (1977) added more dye tracing; cave description and mapping; and geophysical data to reconstruct paleo-flow routes. Mylroie (1984) capped off the earlier studies with a partial review/partial conceptual/partial hypothesis paper on the effects of glaciation on karst, including karst in the Helderberg Plateau, in other glaciated karst areas of the world, and world-wide karst influenced by sea level changes due to glaciation.

The above-mentioned studies provided a partial framework of speleogenesis in the Helderberg Plateau. The speleogenesis for many of the large caves studied was determined to be pre-glacial in origin, with several being interpreted as post-glacial in origin (e.g. Palmer

1972; Mylroie 1977), by these workers. Though the body of fieldwork provided evidence for survival of caves in the northeastern US through multiple glaciations, doubt still existed, only to be erased through further studies (Lauritzen and Mylroie 2000; Mylroie and Mylroie 2004). These studies also laid the way for future investigations, by leaving loose threads, as well as by leaving a predictive framework that allowed the discovery of other large cave systems by cavers through digging efforts (Chap. 5). Of note are leftovers provided by these early studies for future works including unanalyzed cave sediments (only partially described in the early studies), the beginnings of a post-glacial cave model, and the demonstration of post-glacial rebound expanding joints.

3.1.2 Mid 1980s-Early 2000s

Following the early, consecutive, concurrent, and often interrelated studies, scientific work in the Helderberg Plateau, and elsewhere in New York and New England became scattered, both in time and in content. Many of the found, large, pre-glacial cave systems were mostly wrapped up as far as speleogenetic origin by the end of the 1970s. The previous workers of the late 1960s to early 1970s moved onto other karst regions, such as Virginia and West Virginia (Kastning), Mammoth Cave and the Black Hills of South Dakota (Art and Margaret Palmer), and the Bahamas (Mylroie), or moved onto non-karst hydrology and geomorphology (Baker). Several of those who stuck with karst geology moved on to academic positions in non-glaciated areas: Ernst Kastning to Radford University in Virginia, and John Mylroie to Murray State University in Kentucky, followed by Mississippi State University.

Work in the 1980s by Mylroie in the Bahamas demonstrated speleogenesis from inception to large volume caves in the limited time of high sea level, interglacial periods (Mylroie and Carew 1987), confirming the theory (e.g. Palmer 1984) that caves could form to a traversable size in 10,000 years. The time window of 10,000 years could also allow formation of entirely new caves during interglacial time periods in glaciated terrains. Following thoughts on determining post-glacial origins (e.g. Palmer 1972), Mylroie and Carew (1987) presented a conceptual model for determining whether a cave is post-glacial (Fig. 3.2). They noted that numerous infeeders into pre-glacial caves lined up with the current deranged drainage (Mylroie 1977), and that a cave entirely controlled by deranged drainage should be post-glacial.

During subsequent years work was focused on describing the hydrology of some glaciated karst terrains within the northeastern US (e.g. Nardacci 1994), rather than focused on questions pertaining to glaciated karst. Some work did continue however, describing some effects of glaciation on karst. In 1991 the Appalachian Karst Symposium was held in Radford, Virginia hosted by Ernst Kastning. Several studies by Rubin (1991a–c) from the northeastern United States were published in the proceedings of this symposium relating to glaciation and caves, including non-dissolution caves (Rubin 1991a; Chap. 4). Rubin (1991b) discussed morphologies within Clarksville Cave, relating them to glacial meltwaters (Chap. 6). He also related positions

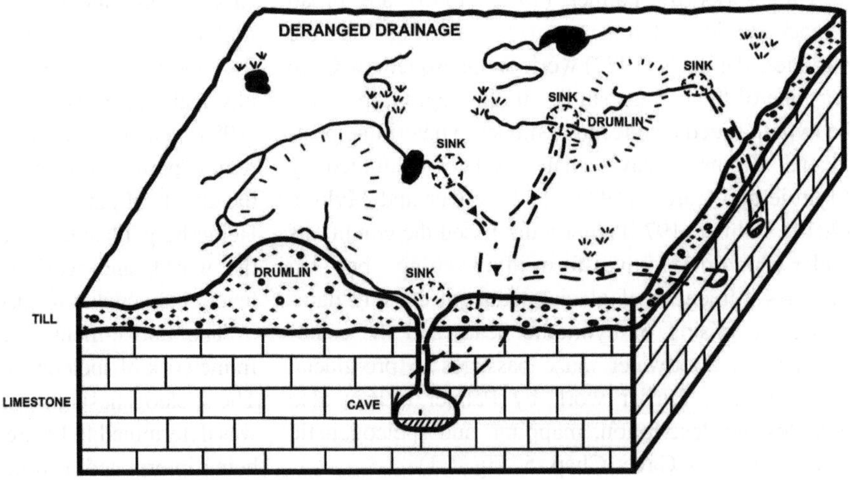

Fig. 3.2 Cartoon drawing of how cave passages, by being in agreement with post-glacial deranged drainage, must therefore be post-glacial in origin. (From Mylroie and Carew 1987)

of littoral cave development to post-glacial rebound in Maine (Rubin 1991a; Chap. 4). These studies by Rubin are some of the first outside of the Helderberg Plateau region, with the littoral cave study being in Maine. Additionally, although Clarksville Cave is located within the Helderberg Plateau, it is in the east, adjacent to the deformed band of Siluro-Devonian carbonates, and has more structural influences than most caves of the Helderberg Plateau (Chaps. 5 and 6). Surficial karren in the northern New York lowlands (Chap. 7) received attention from Tom Feeney (1996), one of the few karren studies done in the northeastern United States.

Two more studies in this time period cemented together missing aspects of the studies of the 1970s. The first follows the discovery of Barrack Zourie Cave in the Helderberg Plateau. This cave was predicted by the work of Mylroie (1977), and was dug open by a group of northeastern cavers in 1992 (Chap. 5). Work by Dumont (1995), a graduate student at Mississippi State University, added data to the flow-routes mapped in earlier studies (Baker 1973, 1976; Kastning 1975; Palmer 1976; Mylroie 1977). Dumont (1995) also performed paleomagnetic studies on cave sediments, and U/Th age dating on speleothems. This study also completed the speleogenetic history of the McFails/Barrack Zourie cave systems and their flow path evolution.

The last study of this time period added the final nail in the coffin to the question of the origin time of the large cave systems in the Helderberg Plateau. Stein-Erik Lauritzen, who had earlier analyzed speleothems in Norway and obtained pre-glacial (pre-Weischelian) U/Th age dates (Lauritzen and Gascoyne 1980; Chap. 1), collaborated with John Mylroie to date speleothems in the Helderberg Plateau (Lauritzen and Mylroie 1996, 2000). Dates obtained in this study reach to the then limit of U/Th age dating, >350 ka, demonstrating that multiple caves in the glaciated northeastern US were pre-glacial in origin (Fig. 3.3).

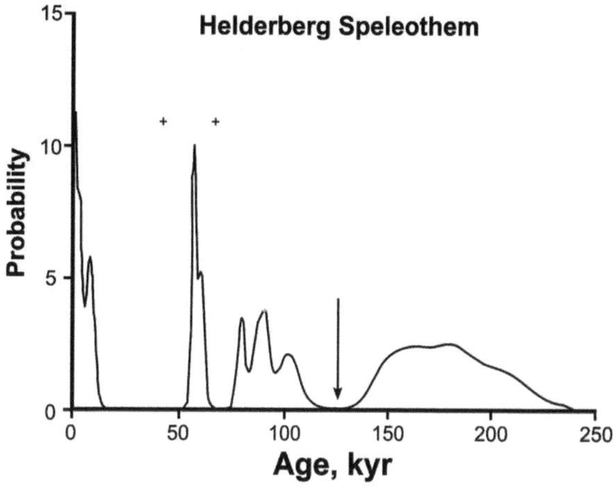

Fig. 3.3 U/Th alpha-count age dates from the Helderberg Plateau and the Hudson River Valley. Ages in excess of 350 ka not shown; the arrow points to an interesting dearth of speleothem growth at 125 ka, during MIS 5e, the last interglacial when this area should have been ice free. The two (+) symbols are U/Th ages from Hollyhock Hollow, south of Onesquethaw Cave, where ice retreat and re-advance around the Wisconsinan interstadial (MIS 3) would have been initiated earlier, and ended later, respectively, than for caves farther north; see Chap. 6 (from Lauritzen and Mylroie 2000)

3.1.3 Early 2000s-Current

The work of Lauritzen and Mylroie (2000) demonstrated once and for all that caves in the northeastern United States, particularly those large systems of the Helderberg Plateau, could survive multiple glacial advances of the Wisconsinan glaciation, as well as previous glaciations. This study gave the final evidence to change the perception of the region as a karst area from the idea that all the caves were, as put by Kastning (Mylroie and Mylroie 2004), "low, wet, muddy, and sparse in formations," to that of having multiple, large, pre-glacial cave systems. Following Lauritzen and Mylroie (2000), a review paper by Mylroie and Mylroie (2004) chronicled the work of the 1960s–2000 on glaciated karst in the Helderberg Plateau, and how it changed the perception of the region, and on glaciated karst in general. Following that review little scientific work other than dye tracing studies in the Helderberg Plateau (e.g. Siemion et al. 2005; Woodell 2004) was conducted either in the Helderberg Plateau, or elsewhere in the northeastern United States until the end of the 2000s decade. Efforts by cavers to locate large caves still continued, such as the 2005 discovery of Thunder Hole in the Helderberg Plateau (Chap. 5), which was aided by the work of Woodell (2004), who demonstrated sink to resurgence connections.

Elsewhere in the world, particularly Norway, earlier and then current efforts on glaciated karst offered hypotheses and evidence for sub-glacial and ice-contact speleogenesis (e.g. Lauritzen 1984), as well for deglacial and post-glacial speleogenesis (e.g. Faulkner 2008). In some cases, both origins are given for the same systems, providing lively debate in the literature on the competing hypotheses. In a search for comparisons to the glaciated karst of the Caledonides (Scotland and Norway), Faulkner (2009) examined marble caves in New England, as well as in the Adirondacks of New York. This study showed the initial return of scientific study in the northeastern US, and not only to the classic Helderberg Plateau locale.

Recent work in the northeastern US, includes research on glacially derived sediments in "caves-as-repositories" studies (van Beynen et al. 2004; Perzan et al. 2014; Weremeichik and Mylroie 2014), as well as following world-wide glaciated karst studies on cave origin times (Cooper and Mylroie 2014). While the large cave systems of the Helderberg Plateau were demonstrated conclusively to be pre-glacial, with some post-glacial infeeders, several loose ends were left by earlier studies. A figure in Mylroie (1984) (Fig. 3.4) illustrated a sequence of sediments seen in multiple caves in Schoharie County, attributing them to stagnant water in ice-covered conditions, though he did not perform a complete sedimentological study to examine this claim. Several caves in the studies of the 1970s-2000 were claimed to be post-glacial (e.g. Palmer 1972; Mylroie 1977), and a model was presented by Mylroie and Carew (1987) for determining if a cave is post-glacial in origin, though no conclusive studies were performed demonstrating this model in the field (Fig. 3.2). Additionally, the earlier studies provided a mechanism for post-glacial speleogenesis in the form of mechanical expansion of joints (e.g. Palmer 1975; Kastning 1975; Baker 1976), used in the glaciated karst studies of Norway (Faulkner 2006), but not fully examined in the northeastern US.

Following the sequence of sediments presented in Mylroie (1984), and local claims that these sediments were glaciolacustrine in origin, a masters student of John Mylroie's at Mississippi State University, Jeremy Weremeichik, analyzed these laminated sediments in several caves of the Helderberg Plateau (Weremeichik and Mylroie 2014). This study revised the interpretation of Mylroie (1984) that these sediments were deposited in stagnant, ice-covered waters to being deposited under the footprint of the hypothesized Glacial Lake Schoharie (Chap. 5, Fig. 5.18). The use of cave sediments in this instance demonstrated survival of deposits in caves, where sufficient surficial deposits are rare and difficult to locate. In a similar "caves-as-repositories" vein, simultaneous work in Vermont by Perzan et al. (2014) analyzed cave sediments, assigning dates of differing packages of sediment to the previous interglacial. While the sediment packages for both of these studies are similar (finely laminated clay and silt), Perzan et al. (2014) did not interpret the sediments in Vermont as glaciolacustrine.

Another recent study closed the loose end of post-glacial speleogenesis, locating field examples of

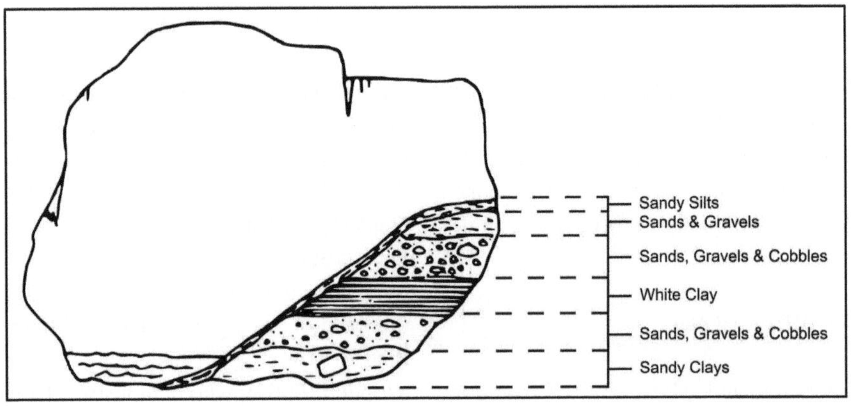

Fig. 3.4 Diagrammatic representation of a sediment sequence from Caboose Cave, Helderberg Plateau, New York (from Mylroie 1984). The sand and gravel transition upward to white carbonate-rich clay and then to sands and gravels again was interpreted to be glacial advance, stillstand, and retreat, respectively. Weremeichik and Mylroie (2014) reinterpreted the clays as glaciolacustrine. See Chap. 5, Fig. 5.18c for a photograph on which this figure was based

post-glacial caves. Cooper and Mylroie (2014) demonstrated that floodwater maze caves in New York are post-glacial, as mechanical enlargement of joints, combined with high hydraulic gradients of proglacial lakes and conditions during deglaciation allowed for rapid dissolutional breakthrough, and floodwaters allowed cave development at maximum rates. The maze caves in this study fit the model of Mylroie and Carew (1987), with the caves being controlled entirely by glacial features, existing in deranged drainage, and only having levels associated with post-glacial (Holocene) base levels.

These recent studies have demonstrated that there is still work to be done on glaciated karst in the northeastern United States. While work in other continentally glaciated karst areas such as Norway has been constantly ongoing, work in New York and New England has come in bursts, followed by hiatuses. As there are still interesting problems to solve in glaciated karst, and problems currently being studied elsewhere on glaciated karst, New York and New England can continue to be an important study area on the subject both compared to other regions, and as a source for new ideas and work.

3.2 What Did These Studies Tell Us About Glaciation and Karst?

As noted above, and in the review paper by Mylroie and Mylroie (2004), the northeastern United States, and in particular the Helderberg Plateau in east central New York has been an important study area in determining the interaction between glaciation and karst. Glaciation imposes various controls on karst (Chap. 1), and work in the northeastern US demonstrated rerouting infeeders and adapting them to post-glacial deranged drainage, lowering or raising base level, clogging of cave systems with glacial sediment forcing backflooding, and removal of surficial and shallow subsurficial karst. Glaciation also imposes important timing controls on karst, where enlargement and deposition of cave minerals typically only occur during short (<20 ka) interglacial, or shorter interstadial periods. Work in the northeastern US has also demonstrated caves as preservational time capsules for sediments and speleothems of past glaciations, where dateable surface materials have been removed.

Work in the Helderberg Plateau and elsewhere in the northeastern United States has played an important role in understanding glaciated karst, especially regarding flow route controls, base level controls, and timing by glaciation. These studies allow a framework to be created understanding the karst landscape evolution through the multiple glacial cycles of the Quaternary.

3.2.1 Flow Route and Base Level Controls

One of the largest contributions of the work in the Helderberg Plateau is the control of glaciation on flow routes. The Helderberg Plateau makes a prime study area for understanding flow route controls by a combination of karst and glacial material as the geology is nearly layer-cake stratigraphy; the stratigraphy of the plateau only contains a shallow 1–2° dip, and few folds or faults (Mylroie 1977, his Table 2; Chaps. 2 and 5). The main structural controls on flow in this region are regional joint patterns associated with the several distant orogenic events (Chap. 2). The relatively small amount of structural control differs from most other locations in the northeastern United States where there is much metamorphism, and/or deformation, which makes teasing out flow route controls more difficult.

Initial studies that included mapping of flow routes are: Baker (1973, 1976), Kastning (1975), Palmer (1976), Mylroie (1977), and Dumont (1995). Each of these studies demonstrated that current, interglacial flow routes take place through established pre-glacial cave systems. These studies have provided a detailed mapping of flow through dye tracing, and described how this flow has been adjusted by glaciation (Chap. 5, Figs. 5.6 and 5.7). The landscape evolution painted by these studies includes the occlusion of previous insurgences and resurgences with glacial sediment. The major flow paths through these pre-glacial caves do not line up with the current surficial drainage regime, and are either pre-Quaternary, or line up with previous interglacials. With the establishment of the current surficial routes, new insurgences form as infeeders to the pre-glacial caves if the current arrangement of drainage and structure allows, rather than entirely new caves as thought previous to the studies of the 1970s. Entirely new caves can also form however, where pre-glacial caves do not exist to capture the flow. New resurgences also form, as occluded pre-glacial resurgences are dammed. Flow routes can then be lowered or raised, particularly due to back-flooding causing overflow routes during high flow (Chap. 5).

Another important result of these studies is the distinction between forms created by karst, and forms created by glaciation. Both of these geomorphic agents create similar features, such as closed depressions. Work such as that by Kastning (1975), has allowed distinction between the two styles, and reinterpreted karstic closed depressions being drained by post-glacial caves, in contrast to closed depressions formed by glacial erosion and damming by moraines, where the pre-glacial cave systems acted as drainage (Chap. 5, Fig. 5.10). Other glacially created depressions may be drained through post-glacial karst systems if the erosion cut below the previous base level of karst development (Cooper and Mylroie 2014). Where pre-glacial base level is higher than these depressions it may be possible to form swamps in them, as the post-glacial cave systems form and are initially inefficient drains. Differentiating between these two styles of drainage becomes important for determining landscape evolution, including previous base levels.

Karstic base levels are also controlled by glaciation. Where glacial erosion removes material base level can lower, either forming new cave levels (Palmer 1976), or entirely new caves (Cooper and Mylroie 2014). Studies in the Helderberg Plateau have also demonstrated a rising of base level due to glacial influence, when tills are deposited in pre-glacial valleys (Palmer 1976; Mylroie 1977; Dumont 1995). Base level rise was determined by geophysical techniques (Palmer 1976; Mylroie 1977), water well log analysis (Palmer 1976), and dye tracing (Baker 1973, 1976; Mylroie 1977) in the Helderberg Plateau, as cave levels abut infilled pre-glacial valleys, and flow upwards through glacial tills such as at Doc Shauls Spring (Chap. 5, Figs. 5.2c and 5.14). Previous, pre-glacial base level positions can also be established through these caves, as development near base level proceeds along strike, whereas above base level cave development occurs down the dip direction. As base level lowers, such as through glacial or fluvial erosion, flow can continue further down dip, as seen in the Helderberg Plateau along the Cobleskill Creek (Chap. 5; Mylroie 1977; Dumont 1995).

3.2.2 Glaciation, Cave Sediments and Speleothems

3.2.2.1 Effects of Glacial Sediments on Caves

Another important aspect on the interaction of glaciation and karst that was demonstrated in the Helderberg Plateau is that of sedimentation. This sedimentation has several possible sources (Mylroie 1977), such as sub-glacial deposition of till, deposition of end moraine material during ice-advance, during deglaciation with large amounts of meltwater mobilizing sediment, or post-glacially as flow resumed through the cave systems (Chap. 5). As mentioned above, flow routes are greatly changed by deposition of sediments occluding pre-glacial insurgences and resurgences (Kastning 1975; Mylroie 1977), as well as by raising base level (Palmer 1976). Occluded resurgences can also produce backflooding, resulting in closed loops and dead end passages in pre-glacial caves. The new, post-glacially adjusted flow routes are partially determined by the routing of water by glacially deposited landforms.

Glacially derived sediments do not only impact karst surface-subsurface interfaces such as insurgences and resurgences. Sediment plugs can additionally form in pre-glacial caves, causing backflooding, similar to plugged resurgences. Art Palmer's seminal maze cave paper (1975) invokes sediment plugs (and other blockages, such as collapse) as important factors in imposing floodwater maze caves in branchwork caves. Palmer (1975) combines this effect with particular attention to glacial sediments, with post-glacial rebound enlarging joints combined to form imposed mazes in glaciated terrains, particularly using the example of Skull Cave in the Helderberg Plateau (Chap. 5, Fig. 5.9).

3.2.2.2 Caves as Repositories for Pre-glacial Materials

The interaction of glaciation and karst does not stop at the glaciers changing pre-glacial karst, and allowing for the formation of new, post-glacial karst. Caves act as important repositories for pre-glacial (pre-Wisconsinan) material, both pre-glacial sediments, and speleothems. As pre-glacial caves survive multiple glaciations, the materials within them also survive this glaciation. This material can then help us understand timing of glacial advance and retreat at higher resolutions for particular areas.

Several studies in the northeastern United States have dated sediments and speleothems within caves (e.g. Dumont 1995; Lauritzen and Mylroie 2000; Perzan et al. 2014). Each of these studies has found pre-Holocene dates for several caves, both in the Helderberg Plateau, and elsewhere in the northeastern US. The dates obtained had direct impact on understanding that caves can survive the multiple glaciations of the Quaternary. These dates may also show certain trends; growth of speleothems tends to only occur during interglacials or during small scale retreats during interstadials, as permafrost blocks water from entering the vadose zone during glacial advance, shutting off speleothem growth (Lauriol et al. 1997) and high hydraulic heads fill in cave systems when glaciers are in place. Lauritzen and Mylroie (2000) performed U/Th age dating on speleothems, demonstrating that speleothem dates can be used to date specific advances and retreats within the Wisconsinan glaciation (Fig. 3.3). While they found dates for small interstadials, they noted that they obtained no dates obtained from the previous interglacial (MIS 5e). They attributed this lack of dates to either sampling bias, or some other factor that did not allow speleothem growth, such as aridity (Lauritzen and Mylroie 2000). The limit of U/Th age dating during the time of this study was 350 ka (using alpha counting), and recent advances in the technique have pushed this time limit to >700 ka. It could be possible that speleothem age dates in the Helderberg Plateau reach back this far, particularly in Schoharie Caverns where dates reached the 350 ka limit.

Perzan et al. (2014) also obtained age dates for sediments in a cave in Vermont, this time by Optically Stimulated Luminescence (OSL). Dates obtained for sands, bracketing rip-up clasts of finely laminated clays, date to before, and after MIS 5e, demonstrating that the rip-up clasts are likely MIS 5e in age, corresponding to high energy flows during the previous interglacial. The finely laminated clay sequences of Perzan et al. (2014) differ from the Helderberg examples (Mylroie 1984; Weremeichik and Mylroie 2014), as they lack large amounts of carbonate material, perhaps a result of limited carbonate outcropping in the western Vermont region.

Cave sediments may also tell us not only about dates of glacial advance and retreat, but also the style of flow or lack of it during glacial and interglacial periods. The presence of these styles of sediments allow interpretations that cannot be made from solely surface data, as surficial landforms are stripped away or extensively modified. A sequence of finely laminated carbonate clays are found in several large, pre-glacial caves of the Helderberg Plateau, within the footprint of the hypothesized Glacial Lake Schoharie, where surficial glacial lake sediments are lacking, likely due to erosion during deglaciation. Weremeichik and Mylroie (2014) used the presence of these finely laminated clays to demonstrate the position of Lake Schoharie, which agreed with surface studies based on ice dam location (Dineen and Hanson 1985). This study also has important implications on the timing of cave development, as certain small caves within the footprint lack these sediments, potentially indicating that the caves did not yet exist at the time when the lake was in place.

The mechanism by which these sediments are emplaced within caves has a great impact on the determination of the age of the cave. The presence of glacial sediments within a cave would seem to indicate the presence of the cave before glaciation; indeed, this may be the case if the sediments were deposited during ice advance, by in-place ice/meltwater, or during deglaciation. Post-glacial caves can contain glacial cave sediments, however, if sediments were transported in through flow, slumping in through entrances, rafted in by organic debris, or if dissolution proceeds upwards to the surface and an overlying deposit of glacial sediments. A paper by van Beynen et al. (2004) demonstrated that pre-glacial caves contain a Holocene speleothem record that revealed aspects of climatic variation over the last 7 ka.

3.2.2.3 Time and Caves in Glaciated Karst Terrains

Another important aspect of glaciation and karst explored in the northeastern United States is the timing of karstification, exploring whether caves are pre- or post-glacial in origin. Prior to the glaciated karst studies in the Helderberg Plateau the interpretation of glaciated karst in continentally glaciated areas of the United States was that caves were mostly post-glacial in origin, with large caves being anomalous (Chap. 1; Mylroie and Mylroie 2004). Contemporaneously with workers in England and Norway (e.g. Waltham 1974; Lauritzen 1981) workers in New York demonstrated that pre-glacial caves could survive multiple glaciations.

The first observation that hinted at pre-glacial origins for these caves was that phreatic, strike-oriented trunk passages were large in cross-section. Though exact rates of cave enlargement were not known to these early workers, the large dimensions at least hinted that these required at least one or more interglacial time periods to form to their current size. The reconstruction of flow routes also indicated pre-glacial ages for caves (e.g. Baker 1976; Mylroie 1977). The large cave systems in the Helderberg Plateau region are currently fed by the deranged drainage. Careful mapping as well as dye tracing indicated that these small infeeders fed into cave systems where there is evidence for previous flow routes that do not line up with the current deranged drainage, nor any deranged drainage that existed in the Holocene. These caves also line up with previous, pre-glacial base levels, further giving evidence of pre-glacial origins (Palmer 1976). The final bit of evidence comes in the form of material dates, in this case U/Th age dates of speleothems (Lauritzen and Mylroie 2000). These age dates reach back past the previous interglacial period, and all the way to the limit of the technique at the time of the study, 350 ka. It may be the case that some of these caves were initiated in pre-Quaternary (Pliocene) times.

The style of some of these pre-glacial caves is branchwork, revealing a dendritic recharge pattern rather than a deranged one. These caves can be long, and have phreatic, strike-oriented passages with large cross-sectional area. The pre-glacial caves have indication of modification by glaciation, including rerouting of flow routes such as by sediment plugging resurgences, and floodwater overprints where glacial sediment dammed flow causing hydraulic heads to rise. These pre-glacial caves, if not relict, also contain adjustments to Holocene deranged drainage, with small infeeders being guided by glacial landforms (Mylroie and Carew 1987).

Observations in the earlier studies also indicated some caves might be post-glacial. In his Onesquethaw Cave study, Palmer (1972) mentioned that this cave aligned with Holocene deranged drainage, and phreatic passage was small in cross-sectional area. Of note in the studies on pre-glacial caves was the post-glacial overprinting on pre-glacial caves with small infeeders aligned to the Holocene drainage regime. Following this, Mylroie and Carew (1987) presented a conceptual model that an entirely post-glacial cave would be controlled by the Holocene deranged drainage regime, with no indication of pre-glacial passage. These caves would additionally be small in cross-section, as the time since deglaciation ($\sim$13 ka for the entirety of New York and New England) could not produce caves with the large cross-sections seen in pre-glacial caves. Additional criteria can also be established to distinguish post-glacial caves from pre-glacial, including material dates, and base level alignments (Table 3.1). Cooper and Mylroie (2014) demonstrated the potential for post-glacial caves by examining floodwater maze caves in New York. Floodwater maze caves form at high rates (Palmer 1991), and are controlled by bedding plane partings or joint partings, which may be enlarged mechanically during glacial rebound. Cooper and Mylroie (2014) also noted that floodwater maze caves are shallow, and therefore are likely to be removed by subsequent glaciation, as opposed to the deep, pre-glacial, branchwork systems. Evidence for these maze caves being post-glacial is that they align with the post-glacial deranged drainage, and to post-glacial base levels, even in the case where the caves are relict, or only receive floodwaters periodically

Table 3.1 Criteria for determination of time origins of caves in glaciated terrains[a]

Criterion	Pre-glacial	Post-glacial
Cross-sectional area	Large, with some small passages	Small
Connection to drainage	Major passages do not reflect Holocene deranged drainage, though small active tributaries can	Directly controlled by deranged drainage, no pre-glacial passage
Base level	Aligned with pre-glacial base level	Aligned with current base level or aligned with previous post-glacial base levels
Glacial sediments	Present, can exist as sequences	Autogenically transported
Material dates	Range from pre-glacial to post-glacial	Will not have age dates older than glacial retreat

[a] While even if all post-glacial criteria are met, it is merely suggestive of post-glacial origins, rather than exact determination: if just one of these criteria fails, it suggests the cave is pre-glacial or sub-glacial in origin; post-glacial over-printing can obscure cave origin

during present-day high discharge events. Additionally, potential formation times were calculated based on cross-sectional area, and the amount of time these caves spend in conduit/pipe-full conditions.

Caves in the northeastern US that have been claimed to be post-glacial in origin have differing geometries in plan and profile views relative to pre-glacial caves. In profile view post-glacial caves are likely to be single level, unless significant Holocene base level changes have occurred in their region. On first intuition it could be thought that in plan view post-glacial caves would be simple, single conduit systems maybe with some tributary passages, as the time required to enlarge passages to humanly enterable sizes would require short breakthrough time, and thus long, branching systems would be difficult to form. Another intuition is that passage cross-sectional areas would be small. These intuitions may be correct for non-floodwater systems such as Westfall Spring Cave of the Helderberg Plateau (Chap. 5, Figs. 5.6 and 5.18). When post-glacial caves are formed by floodwaters however, complex maze geometries can form, as enlargement rates are high and water inundates any available flow path. These maze caves can have passages of large cross-sectional areas, especially when aided by mechanical abrasion that is possible during high flow velocities. Additionally, the high interconnectivity of passages can generate large surveyed lengths, such as Glen Park Labyrinth near Watertown, New York, with over 4 km (2.5 miles) of surveyed passage (Chap. 7, Fig. 7.2).

Glaciation can also add time limitations on caves and karst. While the large, pre-glacial cave systems have clearly been demonstrated to survive glaciation, it may not be the case for all caves, and especially not for surficial karst. Epigene maze caves, both floodwater, and those formed by autogenic recharge through insoluble cap rocks, form in the shallow subsurface (e.g. Palmer 2001). Other post-glacial caves are additionally depth limited, as the shorter breakthrough times to establish a traversable cave during an interglacial timespan may limit the depth caves form to. In at least one maze cave case in the northeastern US, nearby glacially plucked bedrock blocks have a greater dimension than the depth from the surface to the ceiling of caves (Joralemon Park karst, Chap. 6; Cooper and Mylroie 2014). For both of these cases, the original idea that caves would be removed by abrasion and plucking may indeed be the case, and limit the time of these caves to interglacials, unless these caves were protected outside the range of subsequent glaciation, or are located at suitable elevation. It may be possible then, that pre-glacial caves, or caves that are more slowly forming over multiple interglacials, may be biased to preservation over the current humanly enterable, Holocene in origin, caves. Given the large size of these pre-glacial caves, and the limited duration of interglacials, the caves may possibly be pre-Pleistocene in age. The argument also applies to surface karst. Is the interglacial time frame long enough to fully establish a cutter and pinnacle karst surface? In Chap. 5, it is argued perhaps not, and that the currently expressed grike and clint landscape in some karst areas (Fig. 5.12) may carry over from interglacial to interglacial.

3.3 Summary

Work in the northeastern United States on the interplay between glaciation and karst occurred contemporaneously with that in other continentally glaciated areas, Britain and Norway. While in Norway studies were nearly continuous, work in the northeastern US was irregular, coming in several waves. The earlier works in the northeastern United States, particularly in the Helderberg Plateau of east central New York, demonstrated that large, pre-glacial cave systems survived several glacial cycles in the Pleistocene, opposed to earlier ideas that the caves were post-glacial in this area. Following work done in Norway, investigation came back to the northeastern US to compare these caves to those in Norway, and to follow up loose ends left by the earlier studies, including the question of post-glacial caves. The current state of the knowledge on glaciated karst from the northeastern US is that there are a mix of large, pre-glacial cave systems with clear glacial modification, as well as smaller post-glacial caves (although, as the Watertown, New York maze caves demonstrate, post-glacial caves can be extensive).

The studies in the northeastern US also provide a framework for understanding the evolution of these karst systems, by showing how flow routes evolve due to changes by glaciation. Therefore, it is possible to explore the landscape evolution of the variety of caves in the region, not only in the classically studied Helderberg Plateau, but the wide range of geologically differing karst areas within New York and New England. Future cave research in the northeastern United States can accomplish "compare and contrast" examinations of caves in the very differing geologic settings, to isolate those speleogenetic controls due primarily to glaciation. Leading candidates would be deranged drainage, sediment occlusion, abrupt base level modification, and joint activation.

References

Anoynmous (1955) Annual Program Bulletin of the Northeastern Regional Organization of the National Speleological Society, 17p

Baker VR (1973) Geomorphology and hydrology of karst drainage basins and cave channel networks in east central New York. Water Resour Res 9:695–706

Baker VR (1976) Hydrogeology of a cavernous limestone terrane and the hydrochemical mechanisms of its formation, Mohawk River Basin, New York. Emp State Geogram 12:2–65

Berdan JM (1948) Hydrology of limestone terrane in Schoharie County, New York. Am Geophys Union Trans 29:251–253

Berdan JM (1950) The ground-water resources of Schoharie County, New York. NY Water Power Control Comm Bull GW-22:61

Cooper MP, Mylroie JE (2014) Post-glacial speleogenesis: verification of a hypothetical model, and the origins of maze caves in glaciated terrains. Cave Karst Sci 41(2):84–95

Dineen RJ, Hanson EL (1985) Deglaciation of the Middle Mohawk and Sacandaga Valleys, or a tale of two tongues. In: RH Lindemann (ed) Field trip guidebook of the New York State Geological Association 57th annual meeting, New York State Geological Association, pp 250–256

Dumont KA (1995) Karst hydrology and geomorphology of Barrack Zourie cave system, Schoharie County. Bulletin V of the New York Cave Survey, New York

Egemeier SJ (1969) Origin of caves in eastern New York as related to unconfined groundwater flow. Natl Speleol Soc Bull 31:97–111

Faulkner T (2006) Tectonic inception in Caledonide marbles. Acta Carsologica 35(1):7–21

Faulkner T (2008) The top-down, middle-outwards model of cave development in central Scandinavian marbles. Cave Karst Sci 34(1):3–16

Faulkner T (2009) Speleogenesis in New England marble caves. In: Proceedings of the Fifteenth International Speleological Congress, Kerrville, vol 2, pp 855–862

Feeney TP (1996) The role of grikes in limestone pavement formation in northern New York State, USA. In: Fornos JJ, Gines A (eds) Karren landforms. Universitat de les Illes Balears, Palma de Mallorca, pp 53–62

Gurnee R, Hulstrunk A, Gurnee J, Hulstrunk W, Fisher J, Anderson R, Vehslage E, Mueller A, Hawley H, Schoenherr J, Chapman L, Sloane B (1958) Gage caverns. National Speleological Society. Northeastern Regional Organization, Huntsville 8p

Gurnee RH, Anderson R, Mueller AC, Limeres J (1961) Barton hill project—a study of the hydrology of limestone terrain, Schoharie County, New York. Bull Nat Speleol Soc 23:1–30

Hauer PW (1969) Caves of Massachuetts. Speece Productions, Altoona, Pennsylvania, 62p

Kastning EH (1975) Cavern development in the Helderberg Plateau, East-Central New York. Bulletin I of the New York Cave Survey, New York

Lauriol B, Ford DC, Cinq-Mars J, Morris WA (1997) The chronology of speleothem deposition in northern Yukon and its relationship to permafrost. Can J Earth Sci 34:902–911

Lauritzen S-E (1981) Glaciated karst in Norway. In: Proceedings, 8th international speleological congress, vol 2. Bowling Green, pp 410–411

Lauritzen S-E (1984) Evidence of subglacial karstification in Glomdal, Svartisen, Norway. Nor Geogr Tidsskr 38:169–170

Lauritzen S-E, Gascoyne M (1980) The first radiometric dating of Norwegian stalagmites—evidence of pre-Weichselian karst caves. Norsk Geogr Tidsskr 34:77–82

Lauritzen S-E, Mylroie JE (1996) Late Quaternary climatic history of the Helderberg Plateau, New York, USA: preliminary results from U/Th dating of speleothems S-E Lauritzen climate change: the Karst record Karst Waters Institute Charles Town, vol 8788. West Virginia

Lauritzen S-E, Mylroie JE (2000) Results of a speleothem U/Th dating reconnaissance from the Helderberg Plateau, New York. J Cave Karst Stud 62(1):20–26

Mylroie JE (1977) Speleogenesis and karst geomorphology of the Helderberg Plateau, Schoharie County. Bulletin II of the New York Cave Survey, New York

Mylroie JE (1984) Pleistocene climatic variation and cave development. Nor Geogr Tidsskr 38:151–156

Mylroie JE, Carew J (1987) Field evidence for the minimum time for speleogenesis. National Speleological Society Bulletin 49(2):67–72

Mylroie JE, Mylroie JR (2004) Glaciated Karst: how the Helderberg Plateau revised the geologic perception. Northeast Geol Environ Sci 26(1–2):82–92

Nardacci M (1994) The hydrology of the Hannacroix Maze Karst; Albany County, New York. Northeast Caver 25 (3):75–83

Palmer AN (1962) Geologic of the Knox cave system, Albany County, New York. Thesis, Williams College

Palmer AN (1972) Dynamics of a sinking stream system: Onesquethaw Cave, New York. Natl Speleol Soc Bull 34:89–110

Palmer AN (1975) The origin of maze caves. Natl Speleol Soc Bull 34(3):56–76

Palmer AN (1984) Geomorphic interpretation of Karst features. In: LaFleur RA (ed) Groundwater as a geomorphic agent. Allen and Unwin, Boston, pp 173–209

Palmer AN (1991) Origin and morphology of limestone caves. Geol Soc Am Bull 103:1–21

Palmer AN (2001) Dynamics of cave development by allogenic water. Acta Carso 30(2):13–32

Palmer MV (1976) Ground-water flow patterns in limestone solution conduits. Thesis, State University of New York at Oneonta

Perzan Z, Munroe JS, Amidon WH (2014) Investigation of last interglacial sediment in Weybridge Cave, Vermont, USA.

In: Geological Society of America, abstracts with programs, vol 46(6), p 349
Quick PG (2010) Vermont caves: a geological and historical guide. National Speleological Society, Huntsville
Rubin PA (1991a) Emerged sea caves and coastal features as evidence of glacio-isostatic rebound, Mount Desert Island, Maine. In: Kastning EH, Kastning KM (eds) Appalachian Karst: proceedings of the Appalachian Karst symposium, Radford Virginia
Rubin PA (1991b) Modification of preglacial caves by glacial meltwater invasion in east-central New York. In: Kastning EH, Kastning KM (eds) Appalachian Karst: proceedings of the Appalachian Karst symposium, Radford Virginia
Rubin PA (1991c) Flow characteristics and scallop-forming hydraulics within the Mill Pond Karst basin, East-Central New York. In: Kastning EH, Kastning KM (eds) Appalachian Karst: proceedings of the Appalachian Karst symposium, Radford Virginia
Schweiker R, Anderson RR, Van Note P, Jurgens R (1960) Caves of Albany County, NY, vol 8. Northeastern Regional Organization of the National Speleological Society Publication, 26p
Siemion J, Palmer AN, Kraemer TF (2005) Use of quantitative dye racing and water chemistry to characterize karst springs in the Helderberg Plateau, east-central New York. In: Geological Society of America, Abstracts with Programs, vol 37(1), p 68
Van Beynen PE, Schwarcz HP, Ford DC (2004) Holocene climatic variation recorded in a speleothem from McFail's cave, New York. J Cave Karst Stud 66(1):20–27
Waltham AC (1974) Limestone and caves of Northwest England. David and Charles, Newton Abbot, Great Britain
Weremeichik JM, Mylroie JE (2014) Glacial lake Schoharie: an investigative study of glaciolacustrine lithofacies in caves, Helderberg Plateau, Central New York. J Cave Karst Stud 76 (2):127–138
Woodell PA (2004) Hydraulic interpretation of limestone solution conduits from dye tracing and flow analysis, Schoharie County, New York. Thesis, State University of New York at Oneonta

Part II

Cave and Karst Regions of the Northeastern US

4 Pseudokarst and Non-dissolutional Caves

Abstract

Pseudokarst caves form by fluid flow (rheogenic—lava tubes and glacier caves); particle disarticulation (suffusion or piping caves, tafoni); by large-scale rock failure (fracture or crevice caves), and by the results of that failure (talus caves); and by coastal processes (littoral or sea caves). The Northeast contains all these pseudokarst cave types, but crevice caves and talus caves are the dominant type, with sea caves abundant but restricted to rocky coasts. Glaciation, by oversteepening slopes, and by subsequent isostatic rebound, greatly enhances crevice cave and talus cave formation. Glacial eustasy and isostatic rebound form, and then preserve, sea caves. The abundance of massive crystalline rocks in the region creates large fractures, and upon failure, large talus blocks, assisting pseudokarst cave development. These cave types are shallow and surficial, and are likely removed by each ice advance, to be formed again upon ice retreat. Only the lava tube of King Phillips Cave in Connecticut is a definite pre-glacial pseudokarst cave. Crevice caves of over 300 m (1000 ft) in length, e.g. Eagle Cave, Adirondack Mountains, NY or Xanadu Cave, Shawangunks, NY, and talus caves of similar or greater size, e.g. TSOD Cave, Adirondacks, NY or MBDATHS Cave, White Mountains, NH exist along with hundreds of smaller examples.

Pseudokarst is a landscape with features similar to karst, but formed by non-dissolutional processes. Pseudokarst forms in a variety of rock types (including soluble rocks) and through a variety of mechanisms. These terrains can be broken down into several categories: rheogenic (lava flow), glacier, piping, permafrost, talus, and crevice psuedokarsts (Halliday 2007). In addition to these pseudokarst caves are littoral (sea) caves and tafoni (Waterstrat et al. 2010). Unlike karst caves, these do not all result from the subterranean flow of water.

Perhaps the most familiar pseudokarst terrain is rheogenic, existing in locations of recent (geologically) volcanism on volcanic islands and within continental settings as a result of cooling of lava flows. Glacial pseudokarst exists in areas of current glaciation, both alpine and continental. It develops within and under glacial ice, and includes features formed by water movement within the ice, and from melting by geothermal sources. Piping pseudokarst forms from the removal of non- or loosely-compacted particles by groundwater flow, and exist in badlands, within loess

M.P. Cooper and J.E. Mylroie, *Glaciation and Speleogenesis*,
Cave and Karst Systems of the World, DOI 10.1007/978-3-319-16534-9_4,

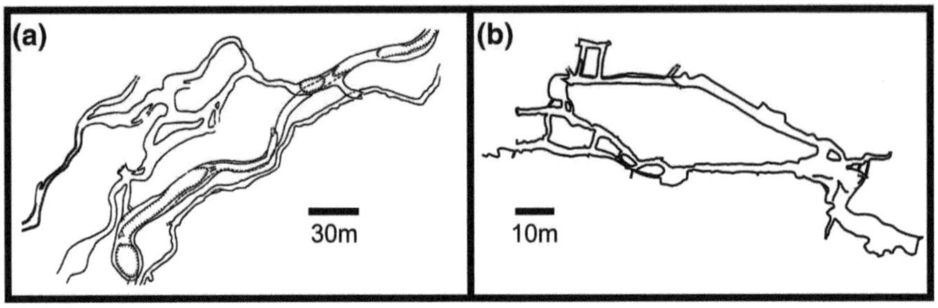

Fig. 4.1 Plan view of a lava tube cave (**a**), and an anastomotic maze cave (**b**). Both of these show similar geometry in plan view, with the appearance of an anastomotic or braided stream. The similar morphology is due to similar process of formation, fluid movement over a planar surface. **a** Sexton portion of Kazamura Cave, Hawaii (modified from Allred and Allred 1997). **b** Big Loop Cave in Essex County, New York (modified from Engel 1989)

deposits, within debris-flow material and within exposed deltas. In 1843, a landslide collapse, at what is now Prospect Park in Troy, New York, formed in an old Glacial Lake Albany delta creating a fissure in the sediment pile that pirated surface water into a short-lived system of piping caves (Porter 1993). Permafrost pseudokarst forms from freeze-thaw cycling and piping, and occurs in areas underlain by permafrost. Talus pseudokarst forms from mass-movement and slope failure, resulting in piles of rock with interstices large enough to admit humans, and as such is restricted to places where large talus blocks form. Crevice pseudokarst forms from mechanical expansion of brittle features such as joints and faults, commonly near cliffs and high slopes. Sea or littoral caves form from the action of waves against rocky coasts, although bioerosion may contribute (see Waterstrat et al. 2010 for a full discussion). Tafoni form from grain-by-grain disintegration of rock of a variety of lithologies; Owen (2013) provides a full review of tafoni.

Pseudokarst terrains may produce caves and can even produce some of the longest and deepest caves in the world, such as the lava tube Kazumura Cave in Hawaii at 65.5 km (40.6 miles) long, and 1.1 km (0.68 miles) deep (Gulden 2014). Some pseudokarst caves can be geometrically and morphometrically similar to dissolutional caves in plan, profile, and cross-sectional views (Fig. 4.1). These similar shapes are generated by similar rheogenic processes such as fluid movement in the case of lava caves, and water flow in glacier caves. These may also have similar speleothems and speleogens such as stalactites in lava tube caves (Fig. 4.2) and scallops in ice caves (Fig. 4.3).

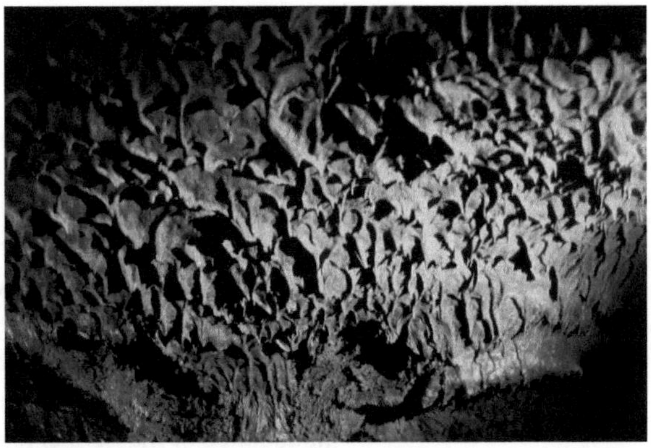

Fig. 4.2 Lava drip forms resembling stalactites, Lava Beds National Monument, California

Processes that cause the initiation and development of dissolutional caves also act on non-soluble rock and can create caves. Tectonics can cause cave inception horizons for dissolution caves (Faulkner 2006), as well as crevice and talus caves (Sjöberg 1987). Enlargement is continued by differing processes, with dissolution for karst caves, and mechanical enlargement by gravitational movement or wave action for crevice caves.

Tectonics influencing cave development, both dissolutional and non-dissolutional is especially prevalent in recently glaciated areas from neotectonics due to isostatic rebound (Sjöberg 1987; Faulkner 2006). In fact, many of the pseudokarstic caves seen in glaciated areas (at least where there is no recent volcanism) may perhaps be a direct result of glaciation and associated neotectonics.

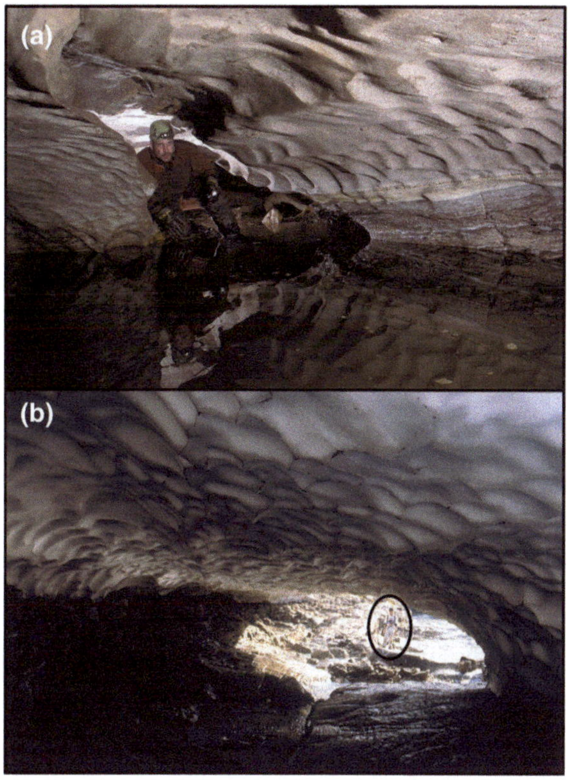

Fig. 4.3 Ablation scallops. **a** Scallops in Cambro-Ordovician marble in a cave stream passage in Shephards Cave, Berkshire County, Massachusetts (*photo* courtesy J. Dunham). **b** Scallops of similar character to those in (**a**) in seasonal ice bridging a stream, Glomdal, Norway [the bedrock floor is Cambro-Ordovician marbles of the same origin as in the (**a**) image]. Air movement and sublimation of the ice forms these scallops. Standing person in circle for scale

4.1 Non-dissolutional Caves and Glaciation

The types of pseudokarstic caves commonly seen in recently glaciated areas (with no volcanism) are talus and crevice caves inland, and littoral (sea) caves along rocky coasts. In some glaciated areas talus caves are the longest caves in their respective regions, compared with dissolutional caves. This includes the New York and New England area, with states such as New Hampshire where the longest caves are in talus, in part because dissolutional caves are non-existent in that state (Chap. 1, Table 1.2).

The most studied region with respect to pseudokarst and glaciation is Scandinavia. Work in Scandinavia has linked glaciation (and in particular neotectonics due to isostatic rebound) to talus caves (Sjöberg 1987, 1996a; Faulkner 2006), crevice caves (Kejonen 1997), and littoral caves (Sjöberg 1988). Pressure release from the retreat of glaciers at the sides of steep slopes, rather than from general isostatic rebound, can also create and enlarge fractures along these slopes to create crevices, and if the slopes fail, talus caves (Lauritzen 1986).

Talus caves, as mentioned before are created from the mass movement of rock from bedrock slope failure. Sjöberg (1987) classified talus caves in Sweden by occurrence in roche moutonées (characterized by smoothed blocks adjacent to roche moutonées), on collapsed mountain slopes, and in displaced mountaintops. Sjöberg (1987) claimed that these talus caves result from neotectonic earthquakes that cause fracturing and slope failure, and has linked them to specific tectonic events dated by disturbed varves (Sjöberg 1996a, b; Faulkner 2006). Neotectonic earthquakes occur from rapid uplift up to 50 cm/year (Mörner 1979), creating tension. This tension follows a curvature that has been estimated by location of talus caves (Ekman 1988), and can be used to further predict their location. These Holocene earthquakes have been estimated to have magnitudes over magnitude eight, with many between six and eight (Mörner et al. 2000). Similar neotectonics have likely occurred in the northeastern United States, as uplift rates are similarly large (Stuiver and Borns 1975; Belknap et al. 1987).

A second factor is that in glaciated regions with significant relief, the production of U-shaped valleys leads to over-steepening of valley walls, and after ice withdrawal, slope failure causes the accumulation of large talus blocks. In both California's Sierra Nevada Mountains, and in the uplands of Norway, the valley incision by glaciers has occurred in massive metamorphic and igneous rocks, resulting in talus blocks of large size. These blocks create interstices that can accommodate human beings. Weaker rocks fail to hold glaciers within steep U-shaped valleys, with subsequent less valley wall over-steepening after ice withdrawal. The subsequent slope failure debris from these weak rocks, where it occurs, is commonly of small dimensions creating voids too small for human entry. In New York and New England, robust and resistant crystalline rocks subjected to glaciation have created over-steepened valley walls, large crevices, and a resultant large talus-block size.

Crevice caves can also be linked to glaciation and the neotectonics following ice retreat. These caves are enlarged brittle features such as fractures and faults, and have regional names such as ice caves or ice gorges (in the northeastern United States), and windy pits in North Yorkshire, Great Britain (Cooper et al. 1976). Initial fractures may have existed prior to the current glacial cycle and are enlarged during, or may be as a result of, the current glacial episode (and thus may not follow regional joint trends; see Lauritzen and Skoglund 2013, their Fig. 3). Crevice caves can be initiated by a variety of processes, including pressure release on the side of glacial valleys (Lauritzen 1986), and by movement of bedrock blocks (Werner and Medville 1995; Murphy and Cordingley 2010), termed mass-movement caves. Further enlargement can be due to gravity, and tectonics. Kejonen (1997) linked the formation of crevice caves in Finland to neotectonics in the period after deglaciation. In Great Britain much work has been done on the North Yorkshire windy pits (e.g. Cooper et al. 1976; Murphy and Lundberg 2008), as well as mass-movement caves (Murphy and Cordingley 2010). This work is largely exploratory and archaeological, though there have been ties to glaciation through speleothem dating and ice cover mapping. Crevice caves in Clark Reservation State Park, in Syracuse, New York, are developed in the Onondaga Limestone, and are a good example of how a massive and strong rock type, though soluble, can support pseudokarst caves (Fig. 4.4).

A subset of crevice caves known as littoral (sea) caves are formed in part by mechanical enlargement of fractures by wave action. Littoral caves overall are intimately tied to glaciation, not only in glaciated areas but worldwide by glacio-eustatic sea-level changes. By tying up water mass on the continents, glaciers can drop sea level as they are advancing, and raise sea level during ice retreat (Chap.1, Fig. 1.5). The elevation of sea caves worldwide can therefore be said to be controlled partly by glaciation. In the glaciated regions these are additionally controlled in elevation by isostatic effects. Sjöberg (1988) tied these caves to the pre-glacial morphology of Norway's coast by noting their positions above any post-glacial local high sea level, most likely correlating with the previous interglacial where sea levels were 6 m higher than present. Littoral caves can also reveal previous relative sea-level positions in the current interglacial (Rubin 1991). Rubin (1991) observed littoral caves on Mount Desert Island, Maine (discussed below) above current sea level, and correlated them to the Upper Marine Limit. While sea level is currently at the highest for this interglacial, the land position relative to sea level was lower from ice loading, and rose from subsequent glacio-isostatic rebound.

The types of pseudokarst linked to glaciation are formed overall by planes of mechanical weakness, and thus are mechanically weak themselves, with talus and crevice cave being formed by loose blocks. This weakness combined with the shallow or surficial location of the caves makes them prone to removal by future glaciation (Sjöberg 1987). This situation is similar to many dissolutional maze caves in glaciated areas (Cooper and Mylroie 2014), and limits the lifetime of pseudokarstic caves. Both talus and crevice caves are therefore limited to the late-glacial and post-glacial time until the next glacial advance. Though this is the case, some crevice caves in the UK have survived (as indicated by speleothem dates) by being located outside of the ice limit of the most recent (Devensian in the UK) glaciation (Murphy and Lundberg 2008; Murphy and Cordingley 2010). Besides the limitation times for these caves, initiation times can and have been explored. Sjöberg (1996a, b) linked talus cave development to specific tectonic events correlated with deformed varves, while Kejonen (1997) also related crevice caves to tectonic events. Additionally, exact dates have been obtained from speleothems in these caves, indicating the

Fig. 4.4 Clark Reservation State Park, Onondaga County, New York. **a** Foundering blocks of Onondaga Limestone, destabilized by cliff failure; cliff edge is towards the very top of the image. **b** Disarticulating limestone surface near the edge of the cliff, which is in the background, to the upper right. **c** Crevice cave entrance, leading downward, in steps, over 30 m (100 ft). **d** Diagrammatic profile of the cliff. In the late 1970s, a trailer was crushed at the trailer park by a falling boulder. The park blamed blasting at a nearby quarry, but the cause was trying to fit too many trailers into the lot by removing the talus slope, destabilizing the cliff behind. Total relief is 50 m (165 ft). **e** Despite the pseudokarst crevice cave development here, dissolutional process are still active, forming karren. Clipboard for scale in **b**, **c** and **e** is 40 cm long

minimum time of development (Murphy and Lundberg 2008; Murphy and Cordingley 2010). Lauritzen and Skoglund (2013) argue that the tectonic trigger effect is overstated; they have pointed out that glacial erosion over-steepens valley walls and leads to joint development parallel to valley walls. As is demonstrated by locations such as Stone Mountain, Georgia, crystalline rocks will produce exfoliation cracks when glacio-isostatic rebound is not a factor. Pseudokarst caves, such as lava tubes, are rare in the northeastern US, but examples are present (Gottlieb 1989) which provides a case of pseudokarst caves that have survived multiple glaciations in the region.

While the majority of studies linking pseudokarst development and glaciation have occurred outside of the northeastern United States, there have been several works on littoral caves in Maine by Rubin (1991), and on crevice caves in New York by Werner and Medville (1995). While there have been not many studies performed here, there has been much work in finding and exploring talus and crevice caves outlined in *The Northeastern Caver*, by the late Robert Carroll, Steve Higham, and others. Typical crevice cave passages are shown in Fig. 4.5.

4.2 Non-dissolutional Caves of New York and New England

In some of the states of the northeastern USA pseudokarst caves are some of the longest caves in the state. Partially for this reason there has been much effort in finding and mapping pseudokarst (particularly talus) caves in the Northeast. These efforts have found many talus caves over 300 m (1000 ft) length (Chap. 1, Table 1.2). The pseudokarst caves in the northeast form in a variety of rock types that range all the groups. Talus caves have been found in anorthosite, granite, gneiss, schist, slate, phyllite, marble, conglomerate, and sandstone (Carroll 1990; Millet and Boop 2013), with crevice caves spanning the same rock types, including limestone. Littoral caves are mostly found in granites along the coast at locations such as Mount Desert Island, Maine, although other

Fig. 4.5 Examples of northeastern crevice cave passages. **a** Cave of the Winds, Lamoille County, Vermont, a crevice cave with cavers descending on vertical gear. **b** W Mountain Cave, Franklin County, New York, with a slanting fissure and ice formations. **c** Pittsford Ice Cave, Rutland County, Vermont showing fissure and breakdown. **d** High fissure in W Mountain Cave (photos courtesy of J. Dunham)

rock types are reported for littoral caves in Rhode Island (Moore 2007) and Vermont (Quick 2012). Northeastern USA pseudokarst caves are under-reported and under-described in the professional literature, and this chapter will remedy that deficiency to some degree by describing a larger set of caves than is done in the succeeding chapters that present dissolutional cave settings.

Perhaps the most prolific mapper of, explorer of, and writer on non-dissolutional caves in the northeast was Robert Carroll Jr., who wrote several dozen articles on them in *The Northeastern Caver* (Higham 2013a). He also mapped, explored, and wrote about dissolution caves in marbles in Vermont and in the Adirondacks of New York, as well as in the Cambro-Ordovician limestones in northern New York. Though the articles Robert Carroll wrote were mainly on exploration, and finding new caves, he was one of the earliest investigators to mention the role tectonics play in creating non-dissolution caves (Carroll 1972), as well as the destructive power of glaciers in regards to non-dissolution caves. He also noted useful parameters for determining where non-dissolutional caves will appear. Through his efforts he located many talus caves, some over 1000 ft (300 m) long (Carroll 1990), and pushed others such as TSOD Cave (Fig. 4.6) to nearly 2.5 miles ($\sim$4 km) long. His work spanned multiple states, where some of the longest caves are non-dissolutional (such as New Hampshire). Some of the caves he found and mapped will be discussed here, as well as non-dissolutional caves found by others such as Steve Higham.

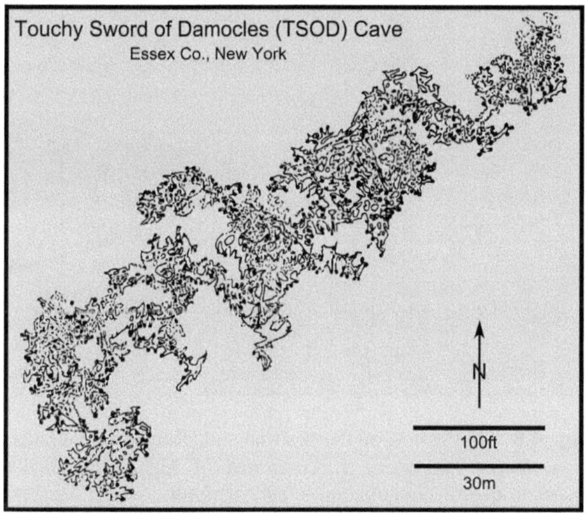

Fig. 4.6 Map of Touchy Sword of Damocles (TSOD) cave. This cave is claimed to be the longest talus cave in the world, though due to the difficulty of defining actual passage in talus caves, the claim of 2.5 miles (4 km) of passage may be doubted. Original cartography of this cave was performed by Robert Carroll Jr. (modified from Nardacci 1991)

4.2.1 New York

Of the states in the northeast, New York has the most, and longest dissolutional caves (Gulden 2014; Higham 2013b) (Chap. 1, Table 1.2). New York is also the home to the longest talus cave in the northeast, TSOD Cave (Touchy Sword of Damocles). Additionally, there are many other non-dissolution caves (including other lengthy ones), in the igneous and metamorphic rocks of the Adirondacks, as well as in the conglomerates and sandstones of the Shawangunks along the west bank of the Hudson River, and the previously mentioned limestone example from Syracuse.

4.2.2 Adirondacks

The Adirondacks of New York are home to the largest non-dissolutional caves in the northeast. This area is a recently uplifted mountain range consisting of Precambrian metamorphic and igneous rocks (Chaps. 2 and 9). Both talus and crevice pseudokarst is developed here, in anorthosite and gneiss. Like elsewhere in the northeast, many of these caves were found and explored by Robert Carroll. Lengthy caves in this region are located in Hamilton and Essex Counties (Carroll 1990), due to high relief and the existence of the Mt. Marcy Anorthosite, the largest body of anorthosite in the United States (Isachsen et al. 2000). The establishment of these lengthy caves is due to the combination of the oversteepened slopes due to glaciation, and the continued uplift of the Adirondacks themselves. Particular talus caves of the Adirondacks are TSOD, Sphagnum Ravine, W.H. Lyman, and Manitou Abode, all claimed to be over 1000 ft (305 m) in length, and within the anorthosite of Essex County. One crevice cave over 1000 ft (305 m) long in the Adirondacks is Eagle Cave, in gneiss of Hamilton County.

TSOD (Touchy Sword of Damocles) Cave (Fig. 4.6) is the longest talus cave in the world (Nardacci 1991) at a claimed 2.5 miles (4 km) of mapped passage and 170 ft (52 m) of relief. It is located in the anorthosite of Essex County, near Indian Pass and the high peaks of the Adirondacks (Nardacci 1991). While some northeastern cavers have sustained this claim, others argue that this cave is many smaller caves that are "linked" in uncovered portions (e.g. Carroll 1997; Halliday 2004). The linked caves include: Old TSOD, TSOD II, TSOD-2B, Ponor, and Strungout caves (Carroll 1977, 1997), found by exploration during the 1970s by Robert Carroll and Roger Bartholomew. Due to these false links TSOD Cave and many other talus caves have been separated into smaller caves upon re-inspection of the cave data (Carroll 1997).

W.H. Lyman Cave is also located near Indian Pass and is also in anorthosite talus. This cave was found by Robert Carroll in 1975, and initially named Henodoawda Cave (Carroll 1978). The total length of this cave is 2500 ft (760 m), formed by connecting Henodoawda I-III caves. In 1978 Robert Carroll renamed this cave to W.H. Lyman Memorial Cave.

Eagle Cave (Fig. 4.7) is a crevice cave in Hamilton County and is formed in gneiss. This cave was found and named by Roger V. Bartholomew in the 1960s (Bartholomew 1987). Additional exploration done by Bartholomew and Carroll brought the total length of Eagle Cave to 1770 ft (540 m) (Carroll 1997); relief to 150 ft (45 m). Eagle Cave, like other crevice caves, formally contained ice well past the winter months. The presence of ice in crevice caves has lent some to name various crevice caves of "ice caves", such as those in the sandstones and conglomerates of the Shawankgunks and elsewhere in the northeast. The term "ice caves" refers to caves with ice in them, not caves within ice (termed glacier caves).

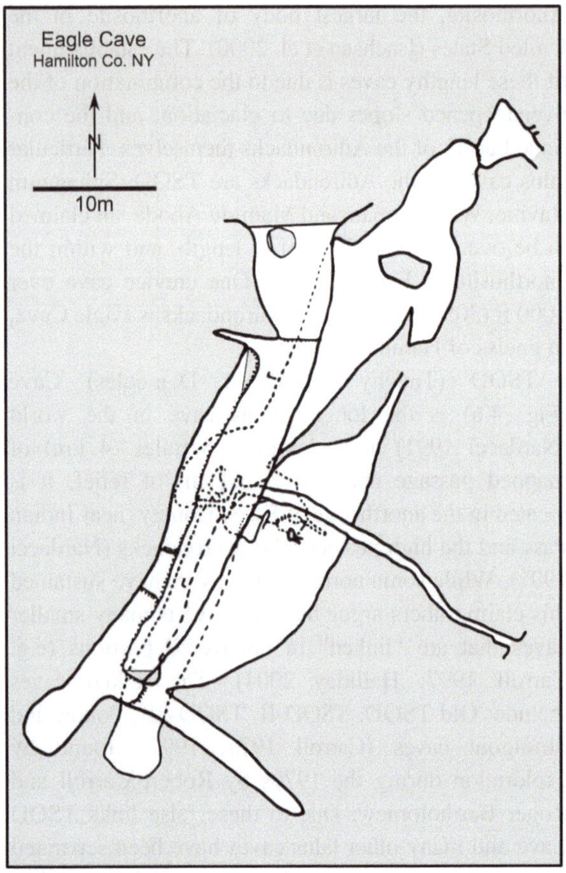

Fig. 4.7 Map of Eagle Cave, a crevice cave formed in gneiss, located in Hamilton County, New York (modified from Bartholomew 1987)

4.2.3 Shawangunks

The Shawankunk Ridge (Fig. 4.8) is also called the Shawangunk Mountains, the Shawangunks, or just "The Gunks". This ridge is mainly composed of quartz pebble conglomerate; however some outcroppings lack the larger grains and are classified as quartz sandstone. The steep, resistant ridge lends itself to development of crevice caves as the uplift and associated neotectonics fractured the pressure-relieved ridge upon ice retreat, accompanied by over-steepened slopes produced by that glaciation. An enlargement of faults has also occurred. This enlargement, combined with mass-movement over underlying shales, enlarges caves to humanly passable dimensions (Werner and Medville 1995).

This area is widely known for recreation, and is particularly famous for rock climbing. Part of this

Fig. 4.8 Photograph of the Shawangunk Ridge in Ellenville, New York. This ridge is composed of highly erosionally resistant quartz conglomerates and sandstones. The resistant nature of the Shawangunk Conglomerate allows the formation of highly oversteepened slopes. As this conglomerate overlies weaker Ordovician shales, sliding may occur due to mass movement and can form crevice caves, and if the slope fails entirely, talus caves (photo courtesy of E. Cooper)

recreation includes hikes that lead to well known and well visited crevice caves such as those of Sam's Point Preserve including and the Sam's Point Ice Caves and nearby Ellenville Fault-Ice Caves. Other crevice caves such as Xanadu Cave exist on and near Mohonk Preserve (Millet and Boop 2013), as well as within Minnewaska State Park such as Slanted Red Rock Cave (Espinasa and Smith 2010).

The Ellenville Fault-Ice Caves are a series of caves on an exposed fault plane. These caves formed from the expansion of the fault, and were roofed over. Like other ice caves, these caves include ice within them well past wintertime. The fault these caves follow extends down to the underlying shale layer, and it has been hypothesized that they formed due to blocks gliding over the shale helped by glacial waters during the retreat of the ice sheet (Werner and Medville 1995). The area around the fault includes several rare plants, and access to these caves is by permit. Similar to the Ellenville Fault-Ice Caves are the other crevice caves of Sam's Point. Unlike the fault caves the initial fractures are due to jointing. These caves also formed by mass-movement and sliding on the underlying shale, and it is likely that if this sliding continues failure would occur and produce talus caves farther down the slope where the blocks come to rest. The caves of Sam's Point, including the Ellenville Fault-Ice Cave have been known at least since the 19th century (Losee 1974). Ellenville Cave is dynamic;

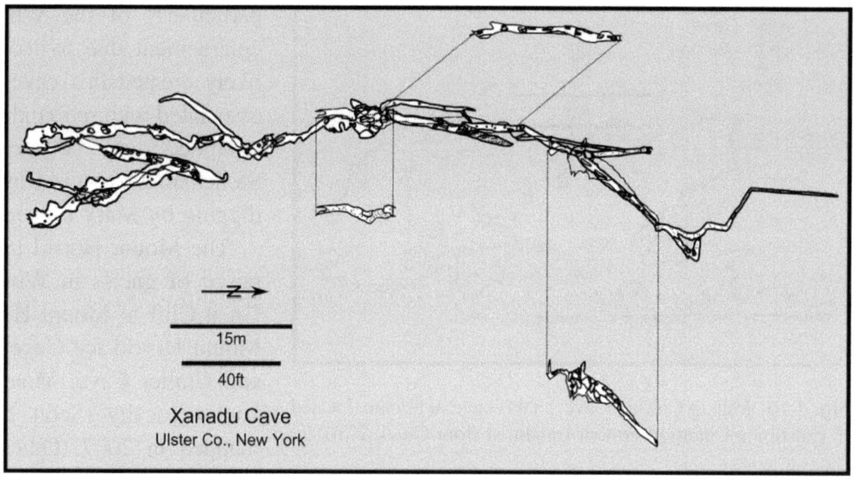

Fig. 4.9 Map of Xanadu Cave, a crevice cave in quartz sandstone, located in Ulster County, New York (modified from Millet and Boop 2013)

its passages changed perceptibly over a five year period. Chockstones disappear and others form new floors, making for very hazardous conditions (C. Porter, pers. comm.).

Crevice caves also exist in the Mohonk Preserve, and Minnewaska State Park. Xanadu Cave (Fig. 4.9) exists near the Mohonk Preserve, and was recently discovered and mapped (Millet and Boop 2013). Xanadu Cave was found and named in 2011 by Bill Griffing, though evidence exists of previous visitation by non-cavers. It quickly became one of the longest and deepest pseudokarst caves in the northeast, with total surveyed length of over 1300 ft (400 m), and 129 ft (40 m) depth. This cave again formed in a similar fashion to the Ellenville Fault-Ice Caves and the ice caves of Sam's Point.

4.2.4 Vermont

After New York, Vermont may have the most cave-related publications of states in the northeastern US. Of most recent note is a book by Quick (2010, 2012) titled *Vermont Caves: A Geologic and Historical Guide*, republished as the guidebook for the 2010 National Speleological Society annual convention in Essex Junction, Vermont. Another, earlier book on Vermont caves is *Caves in Vermont*, by Scott (1959). Both Scott (1959) and Quick (2010, 2012) mention non-dissolution caves, with a section in *Vermont Caves* dedicated to them. As the recent, still available book by Quick (2010, 2012) covers these caves, only a few select caves representing the rocks they form in, and their types (talus and crevice) are presented here. Engel (2000) reports lake-level caves on the New York side of Lake Champlain that appear to be a mix of littoral and dissolution caves, similar features also exist on the Vermont side of the lake (Quick 2012). These caves have formed by the same processes mentioned for other non-dissolution caves in glaciated areas.

Like in New York, Vermont's non-dissolutional caves have formed in a variety of rock types, though dominantly in metamorphic rocks. In Vermont these non-dissolution caves have formed in: granite, schist, gneiss, phyllite, quartzite, and marble. These rocks are mainly Precambrian and Paleozoic, and include Precambrian, Cambrian and Ordovician metamorphic rocks, and Devonian igneous rocks. Both crevice and talus caves are formed in each of these rock types, including some large talus settings that appear like crevices due to their tall, narrow passages (Quick 2010). Of particular note are several non-dissolution caves claimed to be over 1000 ft (305 m) in length: Maidstone Cave, Chiller Cave, Gargantua Cave, and Abenaki Cave (Carroll 1990), though there is some doubt on the claimed lengths due to the mapping techniques used (Quick 2010), and the difficulty in assigning lengths to talus caves (Halliday 2004). Also of note are several talus fields that contain these caves such as the Mount Horrid talus field.

The Devonian granitic rocks contain several caves in Caledonia County, VT. These rocks contain the crevice cave Cow Hill Cave (Fig. 4.10), and the talus cave Abernaki Cave. Cow Hill Cave was known to local residents prior to its description, and was

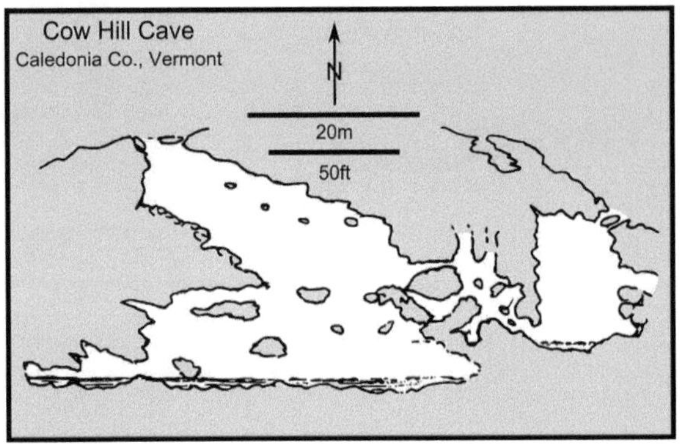

Fig. 4.10 Map of Cowhill Cave, a talus cave in granite, located in Caledonia County, Vermont (modified from Quick 2010)

explored and mapped by Robert Carroll and Miles Drake. This cave has around 450 ft (137 m) of length, and 90 ft (27 m) of relief, and contains near freezing water and ice at its lower reaches. Abernaki Cave was discovered and mapped by Robert Carroll, who claimed it to be 1000 ft (305 m) in length, though revised to 790 ft (240 m) length in updated talus cave-length statistics (Carroll 1997).

Maidstone Cave (Fig. 4.11) is an example of a crevice cave in gneiss of Essex County, VT, on Stoneham Mountain. This cave contains 537 m (1760 ft) of mapped passage (Carroll 1997). It has been known since the 19th century, though it was only mapped relatively recently and described by Folsom and Plante (1995). Plante attributed the cave formation due to initial fracture during Mesozoic plutonism, particularly of the White Mountain Suite, and the enlargement due to isostatic rebound. These factors likely created this cave, along with the neotectonics associated with rebound. Maidstone Cave was initially separated into two caves: Maidstone Cave and Stoneham Mountain Cave, but was connected through digging by Mark Folsom.

The Mount Horrid talus field is a talus field composed of gneiss in Windsor County, adjacent to the Great Cliff of Mount Horrid. This talus field contains Mount Horrid Ice Cave (Fig. 4.12), Gargantua Cave, and Chiller Cave. Mount Horrid Ice Cave has been known locally (Scott 1959), but was only recently mapped in 2007 (Quick 2010) and contains 560 ft (167 m) of mapped, roofed passage. Gargantua and Chiller caves were found by Robert Carroll, and were claimed to have over 300 m (1000 ft) of passage, though some doubt these claims (e.g. Quick 2010).

Non-dissolutional caves of Vermont in mixed schist-quartzite include Widened Fault Cave (as the name suggests, a crevice cave), and Pittsford Ice Cave (talus), both in Rutland County. Widened Fault Cave (Fig. 4.13) is a single passage fault cave mapped to 343 ft (104 m), with a relief of 76 ft (23 m). Pittsford Ice Cave is one of the well-known caves of Vermont (Fig. 4.5c), with some claims of it being known in the late 1700s (Quick 2010). The cave includes ice nearly all year, sometimes persisting into August (Quick 2010). This cave has over 190 m (625 ft) of mapped passage (Higham 2001), and 12 m (40 ft) of relief.

Fig. 4.11 Map of Maidstone Cave, a crevice cave in gneiss, located on Stoneham Mountain, Essex County, Vermont (modified from Quick 2010)

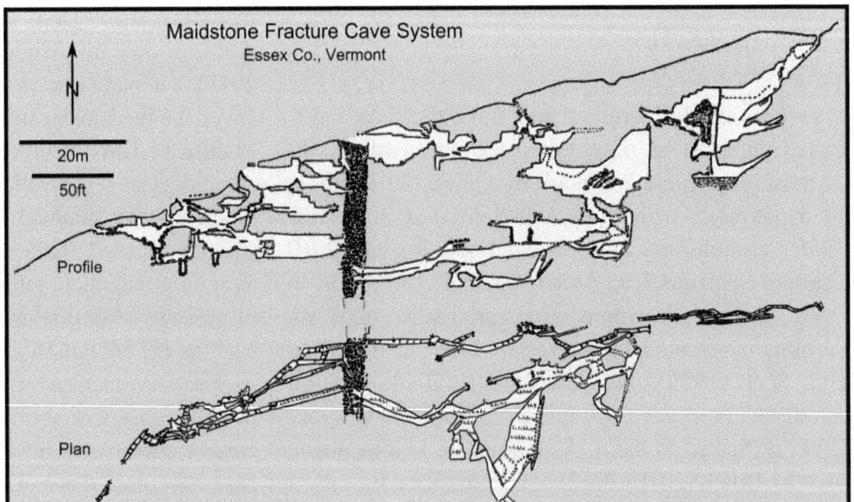

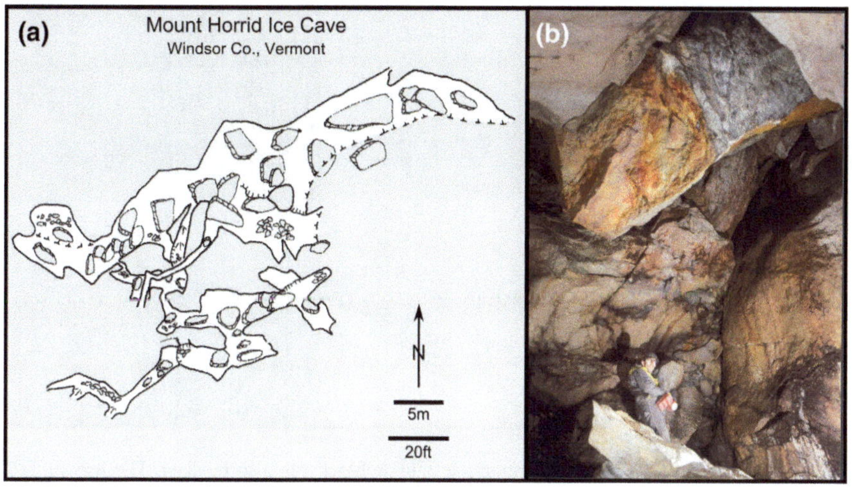

Fig. 4.12 Map of the Mount Horrid Ice Cave (**a**), a talus cave in Winsor County, Vermont. The term "ice cave" refers to the presence of ice well after winter months (modified from Quick 2010). **b** Photograph from Mount Horrid Ice Cave, courtesy C. Porter

Perhaps one of the most interesting non-dissolutional caves in Vermont is Williams Cave (Fig. 4.14), in the schist of Bennington County, named for Frank W. Williams who discovered it in the 1880s. This cave was mentioned in a 1901 USGS Bulletin, attributing its formation due to tension (Quick 2010). Of particular interest is its apparent "bottomless" nature, as it pinches out towards the lower reaches, but continues below in smaller-than-human-sized passage. This cave contains 337 ft (100 m) of mapped passage, and over 150 ft (46 m) of relief (Keogh 2000). Figure 4.14a shows clearly how slope failure released a large glide block, which created crevices that became caves.

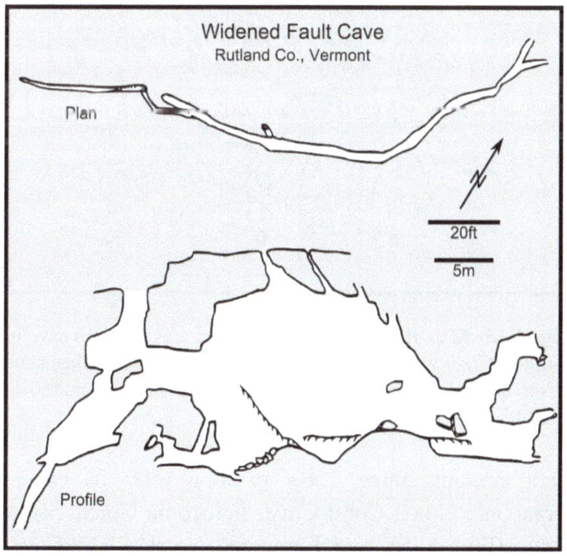

Fig. 4.13 Map of Widened Fault Cave, a crevice cave in Rutland County, Vermont. The passage geometry of this cave is simple, following an enlarged fault (modified from Quick 2010)

4.2.5 New Hampshire

New Hampshire, like the other New England states is geologically composed of plutonic igneous rocks and metamorphic rocks. In New Hampshire soluble rock is fairly limited, and if dissolutional caves existed, they may have been removed or filled in during the multiple glaciations of the Pleistocene. Despite this, there are caves to be found and a good portion of sizeable New Hampshire caves are crevice and talus caves. Efforts, like those in the Adirondacks of New York and Vermont by Robert Carroll and others, have found many talus caves in New Hampshire, some of lengths over 150 m (500 ft) (Carroll 1997). These talus and crevice caves are most commonly found within the plutonic granite, and metamorphic schist and gneiss.

New Hampshire's claim to talus fame is MBDATHS Cave (Merrills-Barn Door-And-The Hole-Scotts) (Carroll 1975), though like other large talus caves has been sliced into multiple smaller talus caves where supposed links were found not to be true links (Carroll 1997). Though now defunct as a complete cave MBDATHS was claimed to be the longest cave in New England at 5300 ft (1615 m) in length (Fig. 4.15). This length was established by connecting smaller talus caves in the granite of Grafton County

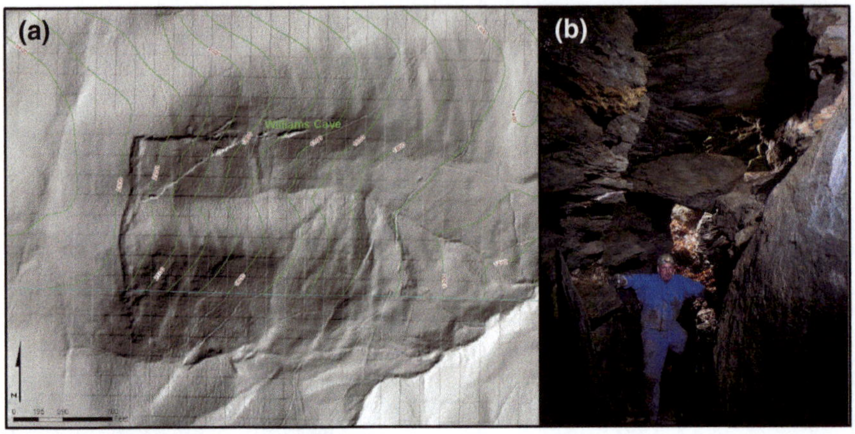

Fig. 4.14 Williams Cave, Bennington County, Vermont, a crevice cave. **a** Lidar image of the area around Williams Cave, with elevations in feet. The block glide is obvious as shown by the main fractures north running through Williams Cave and the right-angle fracture to the west. The compressive land distortion to the east and south of Williams Cave caused by the tow of the glide block can also be seen. The loss of 1100 feet (335 m) in elevation over 3000 linear feet (915 m) creates a 21.5° (37 %) slope, making slope failure possible. **b** Passage in Williams Cave showing fracture origin. (Lidar map and cave photo courtesy J. Dunham)

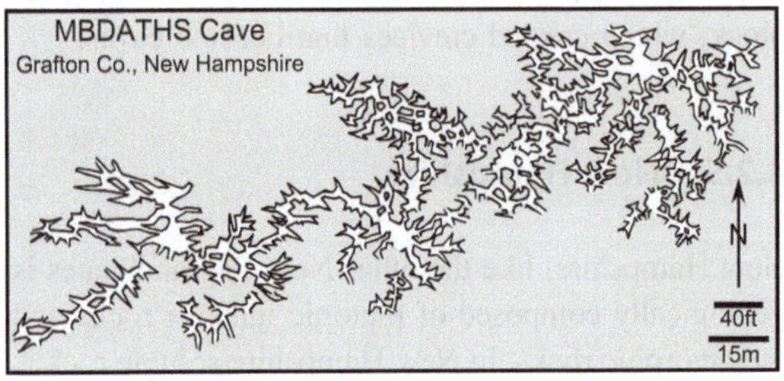

Fig. 4.15 Map of MBDATHS Cave, a talus cave in Grafton County, New Hampshire. This cave exhibits complex network maze geometry. The length of talus caves like MBDATHS may be exaggerated, with tenuous at best links between smaller talus caves (modified from Nardacci 1991)

near Kinsman Notch. The first connection made was between Barn Door and The Hole (BDATH). Additional connections were Merrill's Cave, and Scott's Cave (MBDATHS) through the efforts of Miles Drake, Dave Allured, and Robert Carroll in the mid-1970s (Carroll 1975). Like TSOD Cave in the Adirondacks, the length of MBDATHS Cave was shortened by splitting the cave into multiple caves due to tenuous links (Carroll 1997). While this was the case, the resulting split into Kinsman Talus System West and Kinsman Talus System Scotts caves still contain lengthy passage at greater than 600 m (2000 ft) for each cave.

Another granite talus area in New Hampshire is that of Franconia Notch, also in Grafton County. This talus

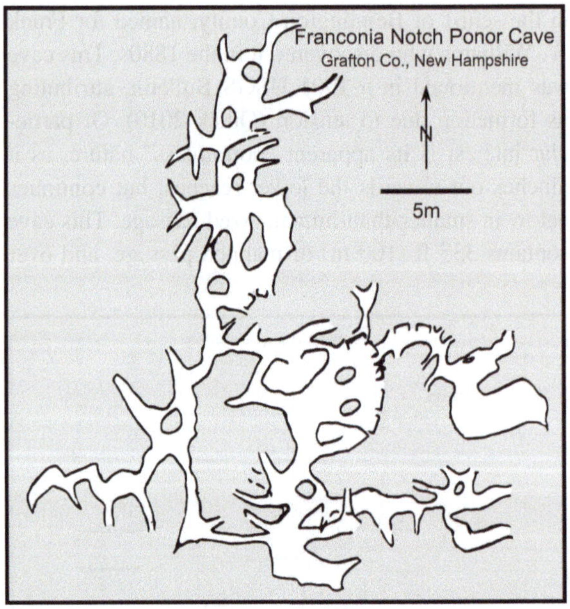

Fig. 4.16 Map of Franconia Notch Ponor Cave, a talus cave in Grafton County, New Hampshire. This and several other talus caves in the Franconia Notch talus field are longer than 150 m (500 ft) (modified from Carroll 1972)

area contains three 150+ m long (500 ft) caves: Franconia Notch Coral Cave, Franconia Notch Ponor Cave (Fig. 4.16), and Franconia Notch Slabs Cave, found by Robert Carroll. Slabs Cave and Coral Cave were found in 1978, and are 180 m (600 ft) length and 240 m (800 ft) length respectively. Ponor Cave was found in 1981 and is 195 m (640 ft) long. Ponor Cave is thought to contain a stream, and contains ice well into the summer months (Carroll 1982).

While the majority of lengthy talus caves in New Hampshire are in granite (Carroll 1997), there do exist some in gneiss. Mount Washington Snow Talus Maze is located on Mount Washington in Coos County, New Hampshire. It is the lengthiest gneiss talus cave in the state at 330 m (1100 ft) in length. Robert Carroll found this cave in 1977 at 1230 m (4000 ft) above sea level, making it one of the highest located talus caves in the northeast. Another gneiss talus cave in New Hampshire is Mt. Adams Ravine Cave, also in Coos County.

In addition to Robert Carroll's efforts, Steve Higham has worked to find non-dissolution caves in New England, particularly in Maine and New Hampshire. An interesting group of non-dissolutional caves found by Higham (2008a, b) occur on Mount Major, in Belknap County. Higham located these caves in 2008. The caves include Mount Major Rockfall Caves #1 and #2, and Mount Major Crevice Caves #1, #2 and #3. These caves are formed in Jurassic aged syenite, an uncommon, quartz-poor, feldspar-abundant igneous rock. While short (all <150 ft, 46 m in length), the occurrence in this uncommon rock makes them a particularly unique set of caves in New England.

4.2.6 Maine

Maine is not only home to talus and crevice caves, it is also home to many wave-action enlarged crevice caves called littoral (sea) caves on its rocky coast (Nardacci 2002). It also has some dissolution caves (Chap. 8). These caves, being coastal caves are additionally affected by glaciation, not only lending their existence to glaciation, but also their position. Their exact position is a result of the interplay of sea level (which is controlled by glacial cycles), and isostatic response to loading and unloading of the crust by the weight of the ice above it (Rubin 1991). The littoral caves in Maine form mainly within the plutonic granites that make up much of the state. Other non-dissolution caves form in the granites, and granodiorites of the state. The large size of Maine, its limited population, and the rugged terrain juxtaposed to poorly drained areas have made locating and exploration of both dissolutional and non-dissolutional caves difficult, and the data base is certainly under-representative.

Mount Desert Island in Maine, the location of Acadia National Park contains several talus, crevice, and littoral caves on its several mountains: Cadillac, Day, and Gorham mountains (Rubin et al. 2002). This island is composed of granite, as well as Siluro-Devonian sedimentary rocks and volcanic tuff, and was formed by glaciers cutting deep valleys, which following Holocene sea-level rise, separated these mountains from the mainland. The littoral caves include: Anemone Cave (Fig. 4.17), Great Head Cave, Stag Cave, Cadillac Cliffs Sea Cave, and Champlain Mountain Sea Cave. These caves were used in one of the few scientific studies relating pseudokarst to glaciation in the northeastern United States (Rubin 1991; Rubin et al. 2002). Of these caves Anemone Cave, Great Head Cave, and Stag Cave are actively forming and exist at current mean sea level (m.s.l.). Cadillac Cliffs Sea Cave and Champlain Mountain Sea Cave both exist above 218 ft (66 m) above m.s.l. The position of these caves was accurately measured and mapped by Paul Rubin and Thom Engel to understand the interplay of isostatic rebound and sea level change during the early Holocene (Rubin 1991). An additional cave in this study is Day Mountain Cave, of indeterminate speleogenetic origin at 440+ ft (133+ m) above m.s.l. This cave appears similar to a littoral cave in pattern, but exists at an elevation high above the accepted amount of known rebound rates which could have emplaced it there. Additional littoral caves on Mount Desert Island include Thunder Hole Cave (Fig. 4.17b), in granite, The Ovens (Fig. 4.18) series of caves in volcanic tuff (Higham 2000), and Bernard Mountain Sea Cave at 235 ft (75 m). As noted by Mylroie and Mylroie (2013), high relief and rocky coastal areas undergo constant over-steepening by wave erosion and debris removal. This action facilitates cliff failure, and the development of crevice caves and talus caves, though both face quick removal by continued erosion. Figure 4.19, their Fig. 1.11 and reproduced here, is especially instructive, displaying a broad, wave-cut bench formed in the last 3000 years; clearly, littoral caves have formed, and been removed, in a geologically rapid sequence. Preservation of sea caves is therefore preferential in glaciated regions where isostatic rebound can bring those caves up and out of the continuous cycle of formation and removal by wave action, thus preserving them.

In addition to the littoral caves of Mount Desert Island, there has been discovery of talus caves. Efforts

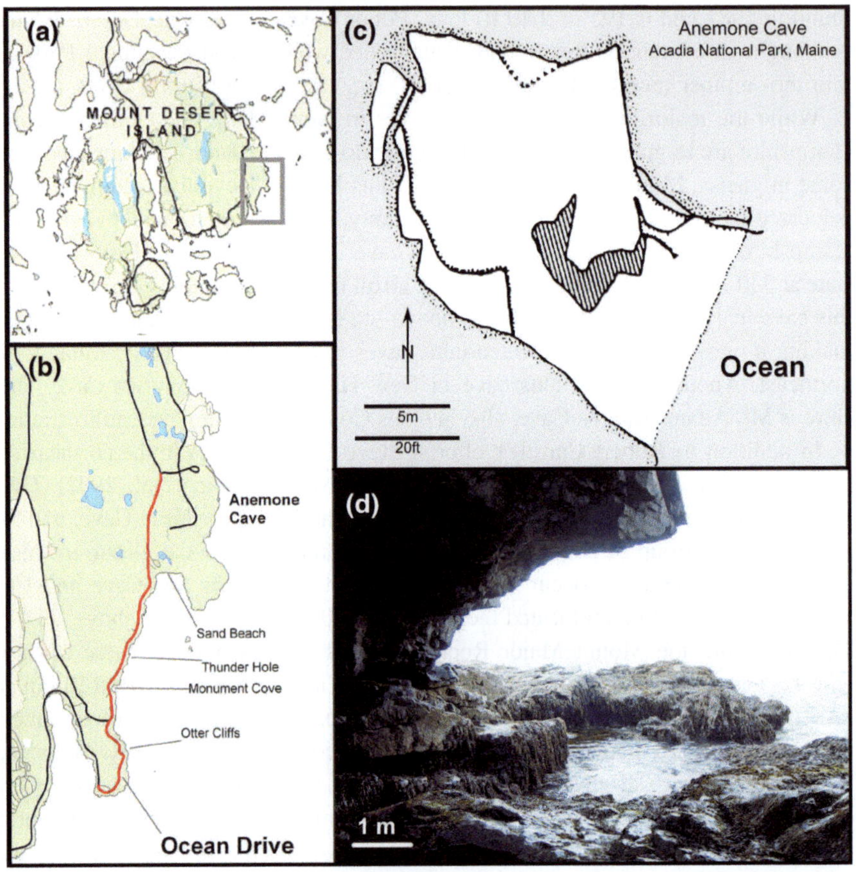

Fig. 4.17 Anemone Cave, Mount Desert Island, Maine. **a** Mount Desert Island, showing location of Anemone Cave (*inset*), and how rugged the island is. **b** Section of coastline hosting Anemone Cave and other sea caves and coastal features. **a** and **b** courtesy of the National Park Service. **c** Map of Anemone Cave (from Rubin 1991). **d** Scene looking outward from inside Anemone Cave (photo courtesy A. Palmer)

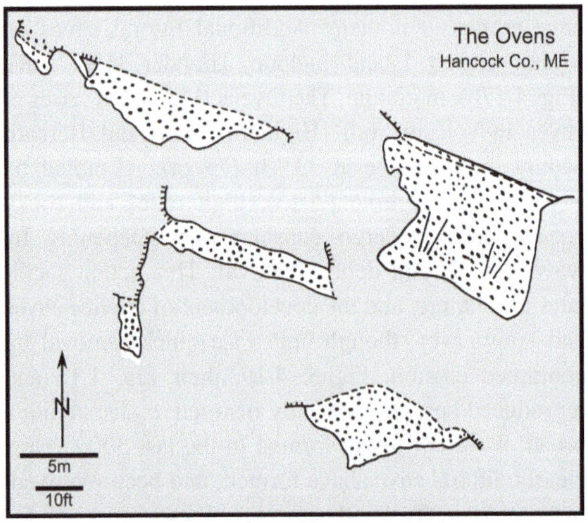

Fig. 4.18 The Ovens are a series of littoral caves developed in volcanic tuff. These caves are fairly unique, as most littoral caves on Mount Desert Island are formed in granite. Original cartography by S. Higham (adapted from Higham 2000)

by Robert Carroll located several talus caves including ones on Beech Mountain such as Champlain Cave, which he reported as having 210 m (690 ft) of passage (Carroll 1983, 1997). Recent work by Morgan Ingalls and John Dunham (Dunham 2010) located several talus caves while searching for ones described by Robert Carroll, on Beech Mountain and Beech Cliffs that they named Enchanted Hall Cave, Underboulder Cave, and Kimball Terrace Hall.

Other efforts by Robert Carroll in Maine produced several 150+ m (500+ ft) long caves, many of which are located in the granites and granodiorites of Piscataquis County. The longest of these is Witherle-ET cave, at 390 m (1280 ft) length making it the longest cave in Maine. Robert Carroll found this cave in Witherle Ravine in 1983, along with several smaller talus caves. In addition to the long talus caves of Piscataquis County is a long crevice cave, the Allagash Ice Cave. It is also the longest crevice cave in

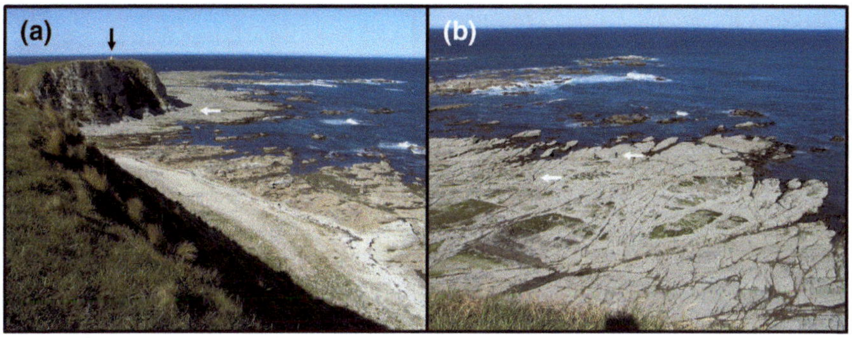

Fig. 4.19 Implications of coastal wave-cut benches. **a** Wave cut bench at the Kaikoura Peninsula, South Island, New Zealand. People at *white arrow* too small to be seen, structure under the *black arrow* is 3 m high. **b** Expanded image of the scene in **a**; *white arrows* point to people barely visible. The entire bench must have been carved since sea level stabilized 3000 years ago; therefore sea caves must have been repeatedly created and destroyed as the sea cliffs migrated landward (from Mylroie and Mylroie 2013)

New England at 300 m (1000 ft). This cave had been known to loggers prior to exploration and mapping by cavers such as Steve Higham and Eric Hendrickson (Higham 1994), and was discovered by loggers in the 1890s (Hendrickson 1997). Although very remote in the Allagash Wilderness, along dirt forestry roads this cave is one of the most visited in Maine. This cave contains ice nearly throughout the year, and has an average temperature of 2.3 °C. Mapping by Steve Higham, and subsequent mapping by Eric Hendrickson and others (Fig. 4.20, Hendrickson 2000) places the cave length at 305 m (1000 ft) with a relief of 23 m (75 ft).

4.2.7 Connecticut and Massachusetts

Connecticut and Massachusetts lack long crevice or talus caves above 150 m (500 ft) in length (Carroll 1997). Though this is the case they contain numerous small talus and crevice caves in the schist and gneiss of the two states. Unlike New York and the rest of New England not much work has been done to specifically locate non-dissolution caves by the likes of Robert Carroll and Steve Higham, as much effort has been spent in locating dissolution caves in marble. This outcome is especially true in Massachusetts, which has more extensive marbles than in the other New England states. Some non-dissolution caves in Connecticut and Massachusetts have received mention in *The Northeastern Caver*, and new talus caves are currently being located (e.g. Girard 2013) in both states.

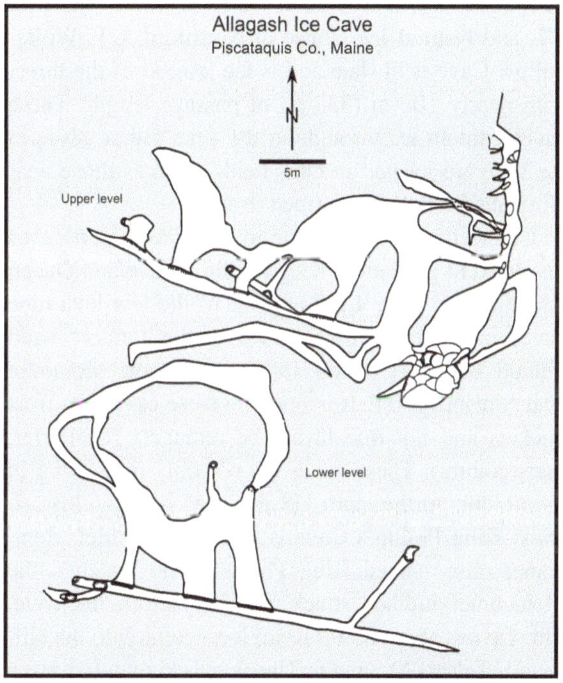

Fig. 4.20 Map of Allagash Ice Caves, a crevice cave located in the remote Allagash Wilderness area. This cave is one of the lengthiest in the state, with 305 m (1000 ft) of passage (modified from Hendrickson 2000)

In Massachusetts, of particular note are the Ice Glen Caves near Stockbridge, Berkshire County, between Bear Mountain and Little Mountain. These have been known since at least the late 19th century (Dunn 1991), and include many small talus caves in granite. Historic accounts of these caves include those written by Perry (1946) in *New England's Buried Treasure*. The Ice Glen contains talus caves at up to 60 ft (18 m) in length, many of which contain ice well into the summer months (Dunn 1991). Another account of the caves at Ice Glen is that of Edwin Balch in *Glacières or Freezing Caverns* (Hauser 1977), who also notes other locations such as Ice Gulf near Great Barrington (also in Berkshire County), and Freezing Well near Ware. One of the best reviews of Massachusetts's non-dissolutional caves by Plante (1990) includes the aforementioned caves, as well as others. Plante (1990, p. 81) suggests some of these caves are "hybrid" dissolutional and crevice within calcareous schists, and dubbed them "tecto-solutional" caves.

Glacières or Freezing Caverns also include Connecticut talus and crevice caves: Wolfshollow Cave near Salisbury, CT, Natural Ice House near Meriden, CT, and Natural Ice House of Northford, CT. Wolfshollow Cave is in slate and is the longest of the three, with nearly 100 m (330 ft) of passage length. These caves contain ice throughout the year. Other caves in the state are located in talus fields, such as those near Norwich (Ash 1990) formed in schist.

Connecticut is also home to non-dissolution caves unrelated to glaciation. King Phillip's Cave and Queen Anne's Cave (Fig. 4.21) are two of the few lava tube caves in the northeast, formed in the Jurassic aged Talcott Basalt (Gottlieb 1989) on Talcott Mountain near Simsbury, CT. It is possible these caves are frost pockets and not true lava tube remnants (C. Porter, pers. comm.). These caves are not only interesting in the unique northeastern US geology, but also historically. King Phillip's Cave is named after chief Metacomet (also named King Phillip), who watched the Metacomet Indians attack Simsbury from the cave. This cave is about 35 ft (10 m) long going into the cliff face of Talcott Mountain. The cave is known to have a variety of minerals, including prehnite (Kastning 1969). The key point for these caves is that if they are lava tubes, then they are obviously pre-glacial, having formed in the Mesozoic, and having survived all the Pleistocene glaciations that reached their latitude. Their opening to the surface, however, may have been the result of recent glacial erosion.

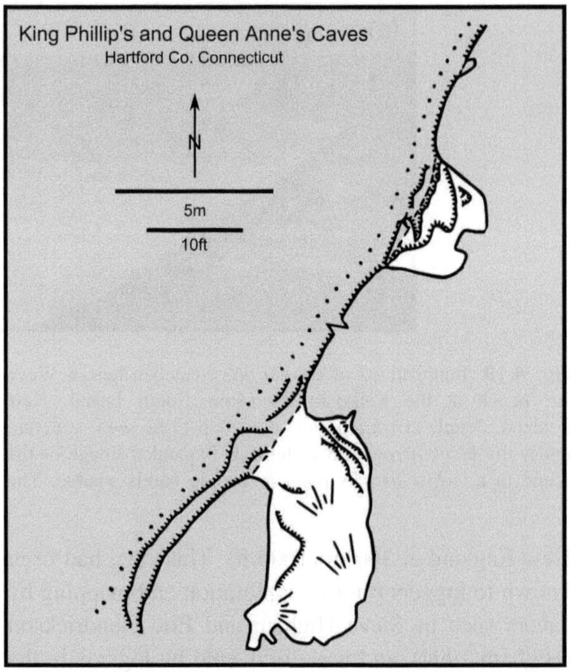

Fig. 4.21 Map of King Phillip's and Queen Anne's caves. These two caves are hypothesized to be Jurassic lava tubes, though other interpretations exist to their origin (modified from Gottlieb 1989)

4.3 Summary

The Wisconsinan glaciation had a direct influence on pseudokarst cave development in New York and New England. The single most important factor is the over-steepening of slopes in massive crystalline rocks, which promotes crevice formation, and subsequently, cliff failure to create extensive collections of large talus blocks. Given that the region's pseudokarst caves are primarily formed in surficial talus deposits, or in cliff-proximal crevices, the Wisconsinan glaciation most likely removed these cave types that had formed from previous glaciations. The caves therefore have an age equivalent to the Pleistocene-Holocene transition or younger. Cave formation continues to this day, as occasional cliff collapse and crevice enlargement still make news headlines, such as the collapse of the Old

Man of the Mountain in New Hampshire in 2003 (Kastning 2008). Sea caves uplifted by glacial isostatic rebound are preserved, and wave action forms new caves at modern sea level. On non-glaciated rocky coasts, the lack of isostatic rebound means that each successive sea cave population is eroded away as the next generation forms. In some rare cases, ancient lava tubes have survived glaciation, but the vast majority of pseudokarst caves in the northeastern United States are less than 15,000 years old.

References

Allred K, Allred C (1997) Development and morphology of Kazumura Cave, Hawaii. J Cave Karst Stud 59(2):67–80

Ash DW (1990) Some Connecticut talus caves near Norwich. Northeast Caver 21(1):31

Bartholomew RV (1987) Eagle Cave discovery. Northeast Caver 18(4):96–98

Belknap DF, Andersen BG, Anderson RS et al (1987) Late Quaternary sea-level changes in Maine. In: Pikey OH, Howard JD (eds) Sea level fluctuation and coastal evolution, vol 41. Society of Economic Paleontologists and Mineralogists, Special Publication, Tulsa, pp 71–85

Carroll RW (1972) Tectonospeleology: a science that needs more attention. Northeast Caver 3(12):144

Carroll RW (1975) MBDATHS: New England's largest cave. Northeast Caver 6(3):74–75

Carroll RW (1977) Diary of an Adirondack talus monster. Northeast Caver 8(5):103–105

Carroll RW (1978) William H. Lyman Memorial Cave. Northeast Caver 9(4):83–85

Carroll RW (1982) Franconia notch Ponor Cave. Northeastern Caver 13(1):24–25

Carroll RW (1983) 1983 Maine double-header: part one: return to the Island. Northeast Caver 24(4):97–100

Carroll RW (1990) 40,000 square mile talus legacy. Northeast Caver 21(4):106–109

Carroll RW (1997) 1997 Reassessment of talus caves and snowtubes. Northeast Caver 28(3):92–93

Cooper MP, Mylroie JE (2014) Post-glacial speleogenesis: verification of a hypothetical model, and the origins of maze caves in glaciated terrains. Cave Karst Sci 41(2):84–95

Cooper RG, Ryder PF, Solman KR (1976) The North Yorkshire windypits: a review. Trans British Cave Res Assoc 3(2):77–94

Dunham J (2010) More talus caves discoveries on Mount Desert Island. Northeast Caver 41:125–126

Dunn R (1991) Ice Glen Caves 100 years later. Northeast Caver 22(3):80–81

Ekman M (1988) Gaussian curvature of postglacial rebound and the discovery of caves created by major earthquakes in Fennoscandia. Geophysica 24(1–2):47–56

Engel TD (1989) Big loop cave. Northeast Caver 20(1):31–35

Engel TD (2000) Caves of Valcour Island, Clinton County, New York. J Cave Karst Stud 62(3):195

Espinasa L, Smith D (2010) Gertrudes nose caves: finding caves where there shouldn't be any. Northeast Caver 41:11–16

Faulkner T (2006) Tectonic inception in Caledonide marbles. Acta Carsologica 35(1):7–21

Folsom M, Plante A (1995) The maidstone fracture cave system. Northeast Caver 26(4):127–141

Girard M (2013) Squaws Kitchen. Northeast Caver 44(4):123–124

Gottlieb MA (1989) Two Connecticut caves. Northeast Caver 20(1):12–14

Gulden B (2014) USA deepest caves. http://www.caverbob.com/usadeep.htm. Accessed 12 Sept 2014

Halliday WR (2004) Talus caves. In: Gunn J (ed) Encyclopedia of cave and karst science. Taylor and Francis, London, pp 1544–1550

Halliday WR (2007) Pseudokarst in the 21st century. J Cave Karst Stud 69(1):103–113

Hauser D (1977) Excerpts from glacieres or freezing caverns by Edwin Swift Balch, 1900. Northeast Caver 8(2):50–51

Hendrickson E (1997) Allagash Ice Caves revisited. Northeast Caver 28(1):22–23

Hendrickson E (2000) Allagash Ice Caves map. Northeast Caver 31(3):96

Higham S (1994) The Allagash Ice Cave. Northeast Caver 25 (4):119–122

Higham S (2000) The Ovens, Mount Desert Island. Northeast Caver 31(4):132–133

Higham S (2001) The Pittsford Ice Cave. Northeast Caver 32 (2):68

Higham S (2008a) New Hampshire cave explorations. Northeast Caver 39:97–98

Higham S (2008b) More talus and crevice caves at mount major. Northeast Caver 39:134–135

Higham S (2013a) The northeastern caver cumulative index: volumes I—XLIV, 1969–2013. http://caves.org/region/nro/neccaver_index.html. Accessed 11 Sept 2014

Higham S (2013b) Northeast long caves, northeast deep caves. Northeast Caver 44(2):44–45

Isachsen YW, Landing E, Lauber JM et al (2000) Geology of New York: a simplified account, 2nd edn. New York State Museum, Albany

Kastning E (1969) Talcott mountain caves. Northeast Caver 1 (6):72–74

Kastning RH (2008) Speleologically-induced demise of the old man of the mountain. Northeast Caver 39(4):115–117

Kejonen A (1997) On Finnish caves. In: Proceedings of the 12th international speleological congress, vol 4, pp 93–98

Keogh J (2000) Williams Cave remapped. Northeast Caver 31 (1):19

Lauritzen S-E (1986) Kvithola at Fauske, northern Norway: an example of ice-contact speleogenesis. Nor Geol Tidsskr 66 (2):153–161

Lauritzen S-E, Skoglund RØ (2013) Glacier ice-contact speleogenesis in marble stripe karst. In: Schroder JF (ed) Treatise geomorphol 6:363–396

Losee P (1974) Structurally controlled caves in the Shawangunk Mountains, Ellenville, NY. Northeast Caver 5(8):91–97

Millet S, Boop L (2013) Xanadu Cave: caverns measureless to man. Northeast Caver 44(2):77–83

Moore J (2007) Early cave compilations of Rhode Island. Northeast Caver 38(1):14–16

Mörner N-A (1979) The Fennoscandian uplift and Late Cenozoic geodynamics: geological evidence. GeoJournal 3 (3):287–318

Mörner N-A, Tröften PE, Sjöberg R et al (2000) Deglacial paleoseismicity in Sweden: the 9663 BP Iggesund event. Quat Sci Rev 19:1461–1468

Murphy PJ, Cordingley JN (2010) Mass movement caves in northern England. Proc Univ Bristol Spelaeol Soc 25 (1):105–112

Murphy PJ, Lundberg J (2008) Uranium series dates from the windy pits of the North York Moors, United Kingdom: implications for late quaternary ice cover and timing of speleogenesis. Earth Surf Proc Land 34:305–313

Mylroie JE, Mylroie JR (2013) Pseudokarst caves in the littoral environment. In: Lace MJ, Mylroie JE (eds) Coastal karst landforms: coastal research library, vol 5. Springer, Dordrecht, pp 3–14

Nardacci M (ed) (1991) Guide to the caves and karst of the northeast. National Speleological Society, Huntsville

Nardacci M (2002) A guide to the geology and caves of the Acadian coast. National Speleological Society 2002 Convention Guidebook Prints Charming, Boothbay, 91 p

Owen AM (2013) Tafoni developmemnmt in the Bahamas. In: Lace MJ, Mylroie JE (eds) Coastal karst landforms: coastal research library, vol 5. Springer, Dordrecht, pp 111–126

Perry CW (1946) New England's buried treasure. Stephen Daye, Brattleboro

Plante AR (1990) Berkshire caves and karsts, Berkshire geology and karst landforms, part-I: history, stratigraphy, and structure. Northeast Caver 21(2):48–56

Porter C (ed) (1993) Catastrophic landslide. Troy Daily Budget in the Northeastern Caver, 24:32

Quick PG (2010) Vermont Caves: a geological and historical guide. National Speleological Society, Huntsville

Quick PG (2012) Vermont Caves: a geologic and historical guide. Paleoflow Press, Jaffrey

Rubin PA (1991) Emerged sea caves and coastal features as evidence of glacio-isostatic rebound, Mount Desert Island, Maine. In: Kastning EH, Kastning KM (eds) Appalachian karst: proceedings of the appalachian karst symposium. Radford

Rubin PA, Engel T, Nardacci M, Morgan B (2002) Geology and paleogeography of Mount Deseret Island and surrounding area, Maine. In: Nardacci M (ed) A guide to the geology and caves of the Acadian coast. National Speleological Society 2002 Convention Guidebook Prints Charming, Boothbay, pp 47–73

Scott JC (1959) Caves in Vermont. Killooleet, Hancock

Sjöberg R (1987) Caves as indicators of neotectonics in Sweden. Z Geomorphol 63:141–148

Sjöberg R (1988) Coastal caves indicating preglacial morphology in Norway. Cave Sci 15(3):99–103

Sjöberg R (1996a) Tavelsjöbergets blockbrant och hur den kan ha bildats. Grottan 31(1):36–40

Sjöberg R (1996b) Lervarv, grottor och jordskalv. Grottan 31 (2):29–31

Stuiver M, Burns HW (1975) Late Quaternary marine invasion in Maine: its chronology and associated crustal movement. Geol Soc Am Bull 86:99–104

Waterstrat WJ, Mylroie JE, Owen AM, Mylroie JR (2010) Coastal caves in Bahamian eolian calcarenites: differentiating between sea caves and flank margin caves using quantitative morphology. J Cave Karst Stud 72:61–74

Werner E, Medville D (1995) The Ellenville (NY) crevice geomorphic processes acting on structural and stratigraphic features. Selected abstracts from the 1995 national speleological society national convention in Blacksburg, Virginia (J Cave Karst Stud)

Caves and Karst of the Helderberg Plateau 5

Abstract

The longest and largest caves of the northeastern region (Barrack Zourie Cave, Howe Caverns, McFails Cave, Skull Cave and Thunder Hole) are found in the Helderberg Plateau of central New York. Broad areal exposure of Helderberg Group limestones, gentle dip SSW at 1°–2°, and down-dip incision of those limestones by surface drainage has created a hydrological setting conducive to large cave formation. Interaction of strike and dip with groundwater release points has created parallel and unconnected down-dip caves above Fox Creek to the east, but integrated and subsequently partitioned strike-oriented caves above Cobleskill Creek to the west. Glaciation has deranged this landscape, particularly in the west where the Cobleskill Creek has been infilled with glacial sediment, backflooding the downstream portions of the McFails and Barrack Zourie cave systems; backflooding has also been severe to the east at Skull Cave, where joint activation due to isostatic rebound has allowed exploitation by floodwaters to create a large rectilinear maze superimposed upon a dendritic cave pattern. A south-flowing tributary valley of Cobleskill Creek has been subdivided by drumlins into a series of large closed depressions that drain to Barrack Zourie and McFails caves. Deranged surface drainage enters many pre-glacial caves by young, post-glacial input routes, and independent small post-glacial caves have developed. Pre-glacial age for the large systems has been demonstrated to be at least 350 ka by U/Th dating, and the perseverance of the pre-glacial drainage pattern through multiple glaciations suggests that the cave hydrology was established pre-Pleistocene.

The caves and karst of the Helderberg Plateau in central New York (Fig. 5.1) are the largest and most well developed in the entire region (Chap. 1, Table 1.1). McFails Cave, at 11.3 km (7 miles) of mapped passage (Fig. 5.2), is the longest in the northeastern United States, north of the Virginias and east of Indiana. While this region includes 12 states, or one-quarter of the 48 contiguous states, it is only 8 % of the area of those contiguous states (624,304 km^2 out of 7,633,941 km^2, or 241,044 miles^2 out of 2,947,468 miles^2). Some of the caves may be the oldest in the northeastern United States, containing stalagmites with ages in excess of 350,000 years (Lauritzen and Mylroie 2000), indicating survival through several complete glacial cycles in the Quaternary.

As noted in Chap. 2, the caves are developed in the Helderberg Group of Late Silurian to Early Devonian age limestones and dolomites, and in the Onondaga

M.P. Cooper and J.E. Mylroie, *Glaciation and Speleogenesis*,
Cave and Karst Systems of the World, DOI 10.1007/978-3-319-16534-9_5,

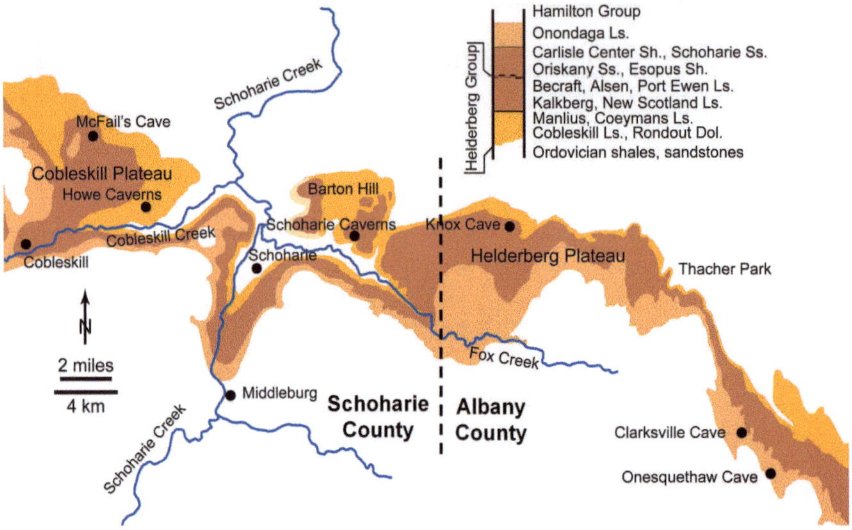

Fig. 5.1 Map and general geology of the Helderberg Plateau in New York State (Engel 2009)

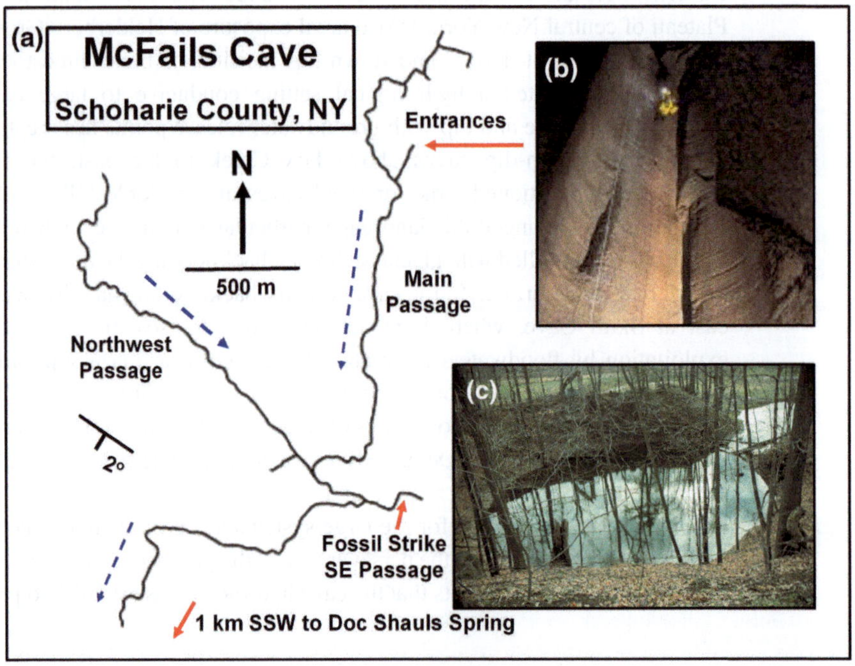

Fig. 5.2 Map of McFails Cave, Schoharie County, New York. **a** Line drawing of McFails Cave, showing strike and dip oriented passages; the fossil Southeast Passage is believed to have once carried water to Howe Caverns (adapted from Palmer 2007). **b** Coeymans Dome in McFails Cave, which leads directly from a surface entrance to a tributary passage and on to the Main Passage (Photo by M. Chu). **c** Doc Shauls Spring, the glacially alluviated resurgence for McFails Cave, water rises ~30 m (100 ft) through glacial till (see Figs. 5.10 and 5.14)

Limestone of Middle Devonian age. Tectonic deformation has been minor, and the rocks dip to the SSW at 1°–2°. However, as the Hudson River Valley is approached on the eastern edge of the Helderberg Plateau, tectonic deformation from the Acadian orogeny has created a complex folded and faulted geologic environment, as addressed in Chap. 6. The limestones of the Helderberg Plateau are sandwiched between thick clastic units that are generally less resistant to erosion than the limestones, resulting in broad limestone surfaces and boundary escarpments up to 30 m ($\sim$100 ft) or more in height (Chap. 1, Fig. 1.7). The limestone plateaus are mantled with a variety of glacial sediments, with till sheets, infilled valleys, and drumlins being the most obvious and abundant.

The position of New York and New England in the northeastern United States has placed them in a location where much early geological research work was done (Fig. 5.3). The early scientific study of caves has been documented from the Helderberg Plateau, with Amos Eaton referring to Clarksville Cave in 1820 in his 2nd edition of *An Index to the Geology of North America*, and with William Mather mentioning caves in his 1843 work *Geology of New York: Part 1* (Engel 2009). Cook (1906) reported on New York caves from the scientific viewpoint; his cave maps were especially well done. For the case of Howe Caverns, his map is the sole remaining scientific documentation of the

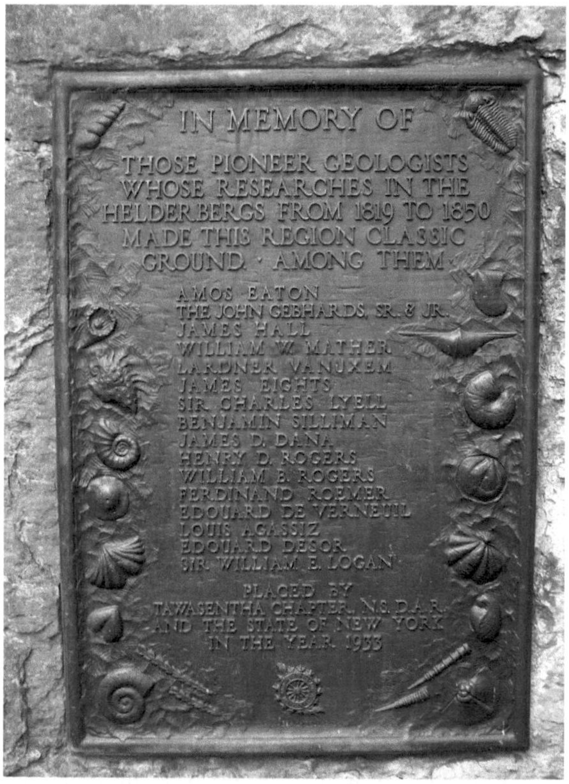

Fig. 5.3 The brass plaque at the western access point to the Indian Ladder Trail at Thacher Park, Albany County, New York. The listed names are those of famous early European and American geologists

downstream portion of that large cave, since quarried away (Fig. 5.4).

Cave and karst areas in flat-lying Paleozoic limestones are common in the United States, mostly from the interior lowland plateaus of states such as Indiana, Kentucky, Tennessee, Alabama and Missouri. The Helderberg Plateau is unique among those states because of glaciation. Minnesota and Iowa also have glaciated karst terrains on flat-lying Paleozoic limestones, but regions such as the "driftless area" of the upper Midwest make those regions different from the Helderberg Plateau, where ice withdrew only $\sim$17–16 ka BP (Ridge 2004). The advantage to studying the effects of glaciation on cave and karst development on the Helderberg Plateau is that the relatively non-deformed nature of the limestones removes a great deal of complexity compared to the Precambrian and Paleozoic marbles of the northeast region, or the highly folded and faulted Paleozoic limestones of the Hudson River Valley.

The Helderberg Plateau compares perhaps best with the Yorkshire Plateau of the United Kingdom, both having relatively undeformed Paleozoic limestones within a clastic sequence. While the Yorkshire limestones are thicker, and have been subjected to a bit more faulting than those of the Helderberg Plateau (e.g. Waltham 1974), the similarities are manifest and both have been repeatedly subjected to continental glaciation during the Quaternary. Another important similarity between the Helderberg Plateau and the Yorkshire Plateau is that they both have been near large population centers for a long time, which has resulted in aggressive and thorough exploration of the caves, creating a cave and karst data base that was begun far earlier, and is much larger, than would be expected if these karst areas were in more remote regions.

The key points are then how has glaciation affected the caves and karst of the Helderberg Plateau? What lessons can we learn from their study? Do the results have broader application than just a compilation of cave maps and descriptions? As noted in Chap. 3, in the 1950s through the 1970s a great deal of cave research was done in the Helderberg Plateau, initiated mostly by post-Second World War recreational cavers but followed up by recreational cavers turned cave scientists. The research work done in the Helderberg Plateau was part of the early nation-wide transition of caving from primarily a recreational activity to a full-fledged scientific discipline.

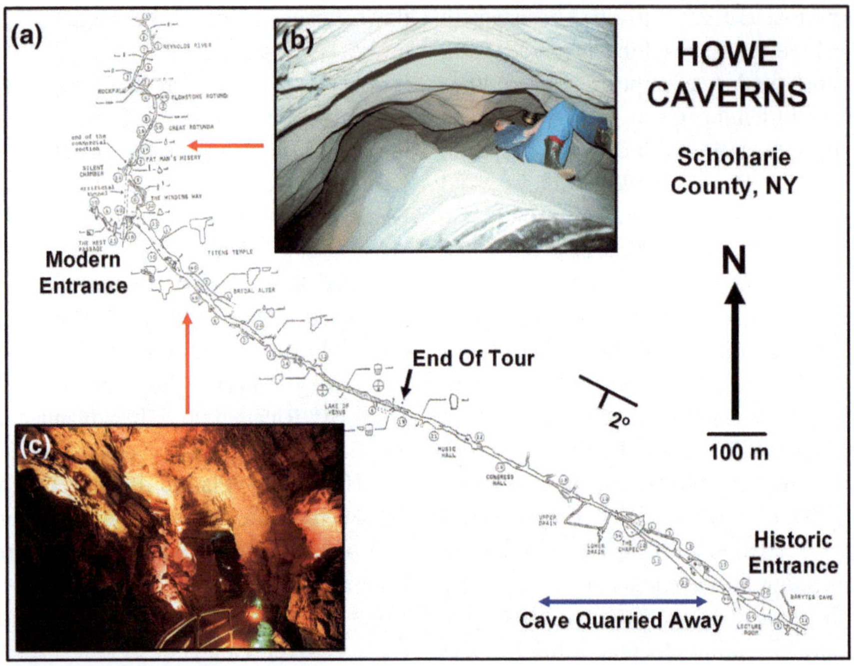

Fig. 5.4 Map of Howe Caverns, Schoharie County, New York. **a** Map of Howe Caverns, highlighting the region quarried away and known only from Cook's (1906) map (adapted from Mylroie 1977). **b** Looking up into the Winding Way, a canyon passage carrying water down dip to the master cave passage. **c** Titan's Temple, looking downstream; a strike oriented master cave passage draining southeast is to the *lower right*, an abandoned upper level to the *upper left* (the colors are due to commercial lighting present in 1976, not dye tracing or pollution)

5.1 Initiation of Modern Cave and Karst Science in the Helderbergs

The initiation of what could be called "modern cave science" in the Helderberg Plateau perhaps began with the Williams College senior B.S. thesis of Art Palmer in 1962, *Geology of the Knox Cave System, Albany County, New York* (Palmer 1962), which created a geologic construct that took into account geology, hydrology and glaciation in a quantitative and scientific way. Art Palmer went on to obtain a PhD at Indiana University and came back in the late 1960s to teach as an assistant professor at Oneonta State University, a short distance south and west of the Helderberg Plateau, where he became the quiet, unassuming, but tremendously influential leader of cave research in New York. Egemeier (1969) published an important but mostly forgotten paper on caves in Albany and Schoharie Counties, and the geochemical, hydrologic and structural controls that affected their speleogenesis. In this paper, Egemeier provided the most complete examination at that time of the major known cave systems, such as McFails Cave, Howe Caverns, and Skull Cave. Another seminal, but often overlooked article by Art Palmer on the Onesquethaw Cave System (Palmer 1972), developed in the Onondaga Limestone in Albany County, helped establish the Helderberg Plateau as an important contributor to cave science. His example motivated others to follow his academic pathway, utilizing cave research to create thesis and dissertation topics that utilized the Helderberg Plateau (e.g. Kastning 1975; Palmer 1976; Baker 1976; Mylroie 1977). This work was influenced by information regarding glaciation and caves coming from the Yorkshire Plateau in the United Kingdom at about the same time (e.g. Waltham 1974).

The main thrust of this early work was to define how the caves systems had functioned as part of the hydrologic framework, applying standard models of cave formation as were understood at the time, and then to see what alterations and differences had occurred that might have been caused by glaciation. The dissertation of Mylroie (1977), having the advantage of drawing on the recent work by Art Palmer (1972, 1975), Kastning (1975), Margaret Palmer (1976), Baker (1973, 1976), summed up the state of knowledge at that time. That dissertation was conducted solely on the Schoharie County portion of the Helderberg Plateau (pronounced *sko-hare-ee*), but as a result, was able to focus on a smaller area to obtain a higher degree of resolution as to how the caves formed, and how glaciation influenced them.

5.2 The General Model of Helderberg Plateau Speleogenesis

A model that attempted to explain speleogenesis in the Helderberg Plateau worked from an initial, non-glaciation viewpoint (Mylroie 1977). The Helderberg Plateau in the Schoharie County region is split into east and west components by the northward-flowing Schoharie Creek, a tributary of the Mohawk River (Fig. 5.5a). Schoharie Creek has incised completely through the leading northern edge of the Helderberg Plateau and its component limestones. Two tributaries to Schoharie Creek, Cobleskill Creek to the west flowing eastward, and Fox Creek to the east, flowing westward, have also incised through the limestones of the Helderberg Plateau (Fig. 5.5b). These stream incisions have created two large blocks of limestone, isolated to the north by the leading edge of the Helderberg Escarpment, to the south by the Cobleskill and Fox Creeks, and from each other by Schoharie Creek, as depicted in Fig. 5.5. The Cobleskill Creek side is called the Cobleskill Plateau; the Fox Creek side is called Barton Hill. The gentle dip of the limestones to the SSW has produced different cave-forming patterns in the two limestone blocks.

Cave genesis began as Schoharie Creek initiated its incision through the limestones, allowing groundwater release from within those units. As Schoharie Creek's flow pattern was approximately perpendicular to the

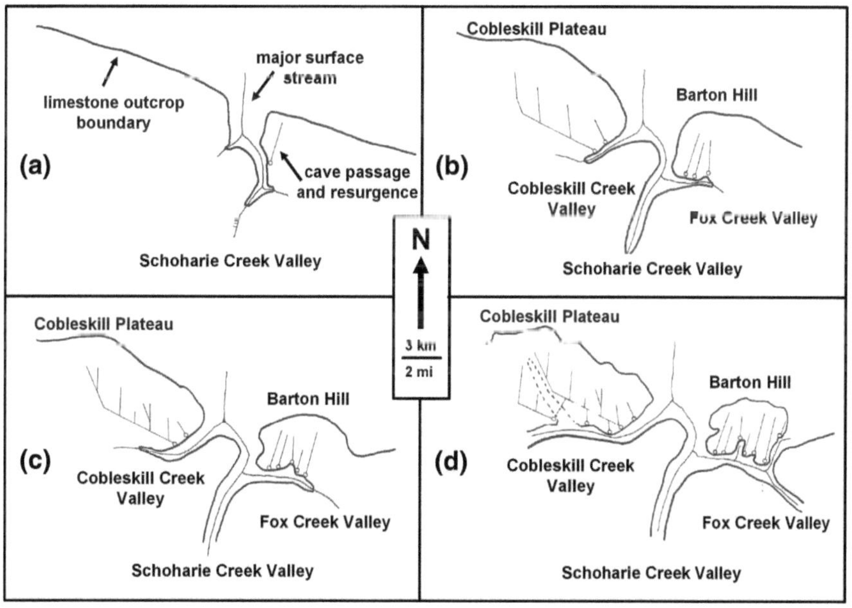

Fig. 5.5 Pre-glacial evolution of the Cobleskill Plateau and Barton Hill cave systems, Schoharie County, New York. See also Figs. 5.6 and 5.7. **a** The initial incision phase of north-flowing Schoharie Creek across the Helderberg Plateau. Cave development is relatively minor. **b** Fox Creek and Cobleskill Creek begin to incise through the Helderberg limestones down-dip and along strike, allowing water release from the Helderberg Plateau. A strike-oriented cave system develops on the Cobleskill Plateau, and down-dip stream caves develop on Barton Hill. **c** The large, integrated cave system on the Cobleskill Plateau continues to enlarge; on Barton Hill, simpler down-dip stream caves are added sequentially to the east as Fox Creek incises eastward. **d** Headward incision of Cobleskill Creek provides down-dip release for the strike-oriented master cave, and it is partitioned into smaller cave segments; a valley begins to incise northward at the west side of the Cobleskill Plateau. A northward incising tributary valley to Fox Creek isolates Barton Hill from the rest of the Helderberg Plateau to the east (location of Thunder Hole and Skull Cave), minor down-dip cave systėms appear on the east side of Barton Hill

strike of the limestones, initial cave development was localized to the immediate area around the Schoharie Creek valley (Fig. 5.5a). Once significant incision had occurred, the limestones in the Cobleskill Plateau began to form a large, integrated cave system that collected water from down-dip flowing tributaries and conducted that flow by a major strike-oriented passage to resurge close to the bank of Schoharie Creek (Fig. 5.5b). At this time, on Barton Hill to the east, cave development was restricted to local down-dip cave passages that discharged to Fox Creek (Fig. 5.5b), although it can be speculated that a large trunk fragment in Gage Caverns (now officially returned to a previous name, Balls Cave, by the National Speleological Society, which owns the cave) with an east-to-west orientation (Fig. 5.6), could be a fossil strike-flow passage (or, alternatively, just a relict meander segment from an abandoned south-flowing stream passage).

When the tributary streams to Schoharie Creek, Cobleskill Creek and Fox Creek, began to incise through the Helderberg limestones, parallel to but significantly down dip from the Helderberg Escarpment's leading edge to the north, down-dip release of groundwater in the Cobleskill Plateau and Barton Hill limestones was possible. In Fox Creek, the headward (eastward) incision of Fox Creek provided progressively more easterly release for down-dip flow paths within Barton Hill (Figs. 5.5b–d). Long, down-dip caves, each independent of their neighbor to the west, developed sequentially eastward as the incision occurred (Fig. 5.6). On the Cobleskill Creek side, a large well-integrated cave system was already in existence, with a major strike passage running from the Northwest Passage in McFails Cave through Howe Caverns to a resurgence just west of the present day junction of Cobleskill Creek and Schoharie Creek (Figs. 5.5b, c). The subsequent headward, westward incision of Cobleskill Creek provided a sequential down-dip release for water in the major strike-oriented trunk passage, and progressive segmentation of that cave system occurred (Figs. 5.5d and 5.7). A large surface valley formed on the western side of the Cobleskill Plateau, draining clastic uplands west of the McFails Cave area (the name McFail's, with the apostrophe, is a common but incorrect usage, both forms appear abundantly in the literature), and incising southward though the limestones to flow into

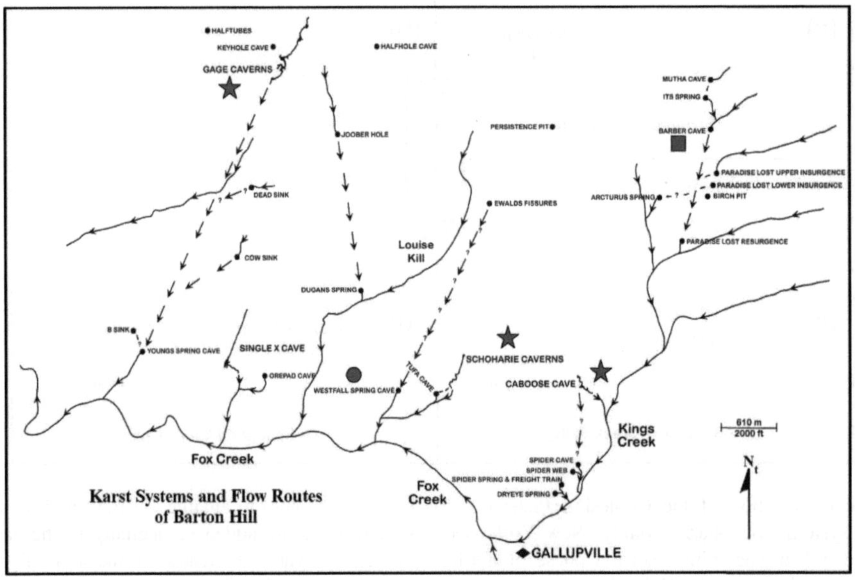

Fig. 5.6 Caves of Barton Hill, Schoharie County, New York. *Stars* indicate caves involved in the glacial sediment study of Weremeichik and Mylroie (2014); the square is Barber Cave, used in the maze cave study of Cooper and Mylroie (2014); the *circle* is Westfall Spring Cave, thought to be post-glacial in origin (Weremeichik and Mylroie 2014). Major caves and their passages named and shown as *solid lines*. Proven dye trace results shown in *solid arrows*, suspected flow connection in *arrows with question marks*. Fox Creek flows along the bottom of the image, *east* to *west*. Thunder Hole lies under the scale and *north arrow* area, Skull Cave is farther east in Albany County

Cobleskill Creek (Figs. 5.5d and 5.7). To the east, Fox Creek continued incising eastward, and a tributary stream also incised to the north, isolating Barton Hill from the rest of the Helderberg Plateau to the east (Fig. 5.5d).

To the west of the Cobleskill Plateau, the limestone outcrop narrows as clastic rock cover is close to the northerly edge of the Helderberg Plateau (Fig. 5.1). Cave development westward into Otsego County is much diminished compared the Cobleskill Plateau, as there is no down-dip release for groundwater in the limestones. Small cave systems draining up-dip to the north dominate. To the east beyond Barton Hill, Fox Creek has several large cave systems on its north bank, first Thunder Hole, a 3 km (1.9 mile) system at the eastern edge of Schoharie County (Fig. 5.8), and then Skull Cave, a 6.1 km (3.8 mile) cave system in western Albany County (Fig. 5.9), with which the aforementioned Knox Cave (Fig. 5.1) is associated. Further eastward, the Helderberg Escarpment begins to make a bend more to the east and south (Fig. 5.1), down-dip incision is lost, and cave development shifts from large, integrated systems to smaller, isolated systems. In this area, cave development in the Becraft Limestone at the top of the Helderberg Group sequence (e.g. Knox Fossil Beckers Cave), and in the Onondaga Limestone in the Middle Devonian sequence, begin to produce caves of intermediate size. Clarksville Cave and slightly eastward, Onesquethaw Cave (Chap. 6), both developed in the Onondaga Limestone, are the largest caves in this eastern edge of the Helderberg Plateau (Fig. 5.1). Clarksville Cave shows strong evidence of control by a thrust fault (Kastning 1975; Rubin 1991a, b), and Onesquethaw Cave displays both faults and low amplitude folds that control its hydrology (Palmer 1972). Further east near the Hudson River, the fold and fault imprint from the Acadian and Alleghenian orogenies is more pronounced and greatly altered speleogenetic patterns (Chap. 6). As Table 1.1 (Chap. 1) demonstrates, the vast majority of long dissolutional caves in the northeastern US are found in the Helderberg Plateau of Albany and Schoharie Counties.

Because Cobleskill Creek and Fox Creek join Schoharie Creek at different positions, Cobleskill joining to the north of Fox Creek, and at an angle to the northeast, a large block of limestone remains between the two, west of the village of Schoharie (Fig. 5.1). This limestone block is called Terrace Mountain, as the Coeymans, Becraft and Onondaga Limestones each make a vertical cliff supporting a terrace or structural bench. Terrace Mountain lacks down-dip incision, and has relatively limited catchment area. As a result, large caves are not found here and Terrace Mountain was ignored by Mylroie's 1977 study. The discovery of

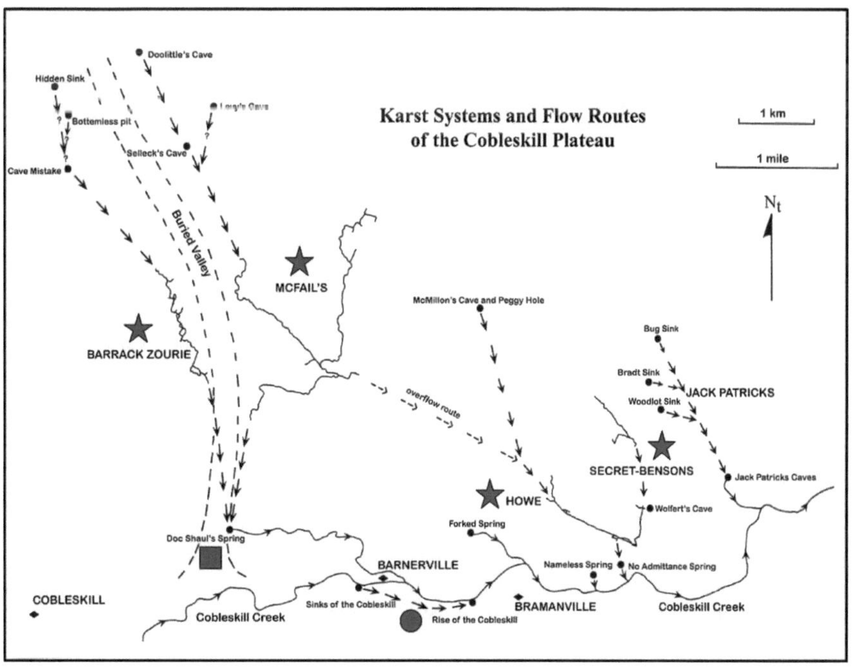

Fig. 5.7 Caves of the Cobleskill Plateau. *Stars* indicate caves involved in the glacial sediment study of Weremeichik and Mylroie (2014); the *square* is Doc Shauls Spring; the *circle* the Sink and Rise of Cobleskill Creek. Major caves and their passages named and shown as *solid lines*. Proven dye trace results shown in *solid arrows*, suspected flow connection in *dashed arrows*. Buried valley shown in *dashed lines*. Cobleskill Creek flows along the bottom of the image, *west* to *east*

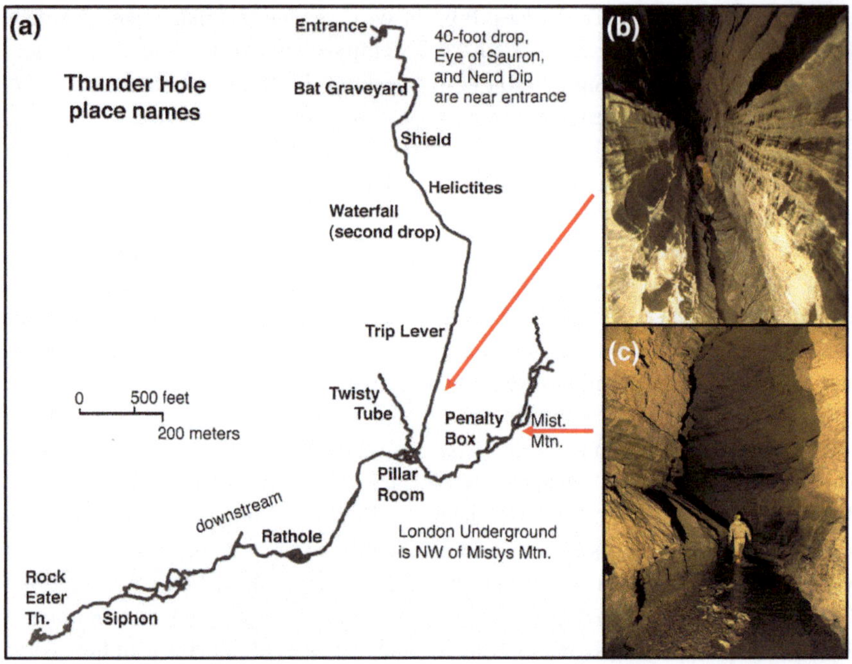

Fig. 5.8 Thunder Hole, Schoharie County, New York. **a** Line map of the cave (from Armstrong et al. 2008). **b** Image of a cave passage developed along a prominent NNE joint. **c** Upstream portion of the main cave, showing passage volumes indicative of a pre-glacial origin. Arrows indicate where photos were taken. (Photos courtesy of M. Chu)

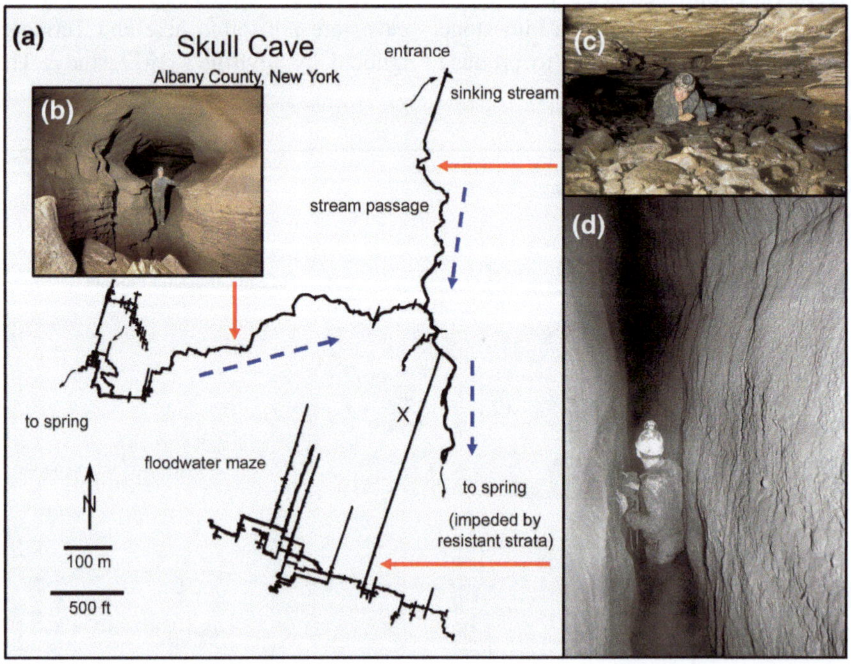

Fig. 5.9 Skull Cave, western Albany County, New York. **a** Line map of the cave (from Palmer 2007); note the converging dendritic stream pattern overlain by a backflood maze utilizing joints opened by isostatic rebound. **b** phreatic tube with a vadose incision. **c** Entrance passage with very coarse glacial sediments of diverse provenances. **d** Long fissure passage (600 m or 2000 ft long) produced by backflooding caused by glacial sediment occlusion of the original resurgence. (Photos courtesy of A. Palmer)

Ain't No Catchment Cave (aka ANC Cave), the longest cave on Terrace Mountain (Armstrong et al. 2005) at 2500 ft (760 m), was a jibe made by cavers about the catchment interpretation (it is the longest cave on Terrace Mountain, but the passages are small and very wet, possibly post-glacial in origin). Veenfleits Cave (aka Van Vleits Cave) is another cave with some complexity, draining down dip and then along strike on the eastern side of the Mountain, directly west of the village of Schoharie. Other caves on Terrace Mountain, such as Lasells Hellhole and Sitzers Cave, are more of historical importance. The most complete scientific work on Terrace Mountain was by Egemeier (1969), who described the effects of faulting and structure on Veenfleits Cave. Terrace Mountain is also involved in the legend of Lester Howe's "Garden of Eden Cave", a mythical cave believed to have been found by Lester Howe after he was bought out of his holdings in Howes Cave (now Howe Caverns), which he had discovered in 1842 (Halliday 1976).

5.2.1 Glacial Overprint of the Helderberg Plateau

The general model for cave formation in the Helderberg Plateau presented above describes speleogenesis in a very traditional way, with down-dip flow either to resurgences, or to a master strike-oriented conduit that delivered water to a resurgence. The maturation of the system, as surface stream incision continued, modified that basic cave plan in a predictable way (Fig. 5.5). Subsequent glaciation overprinted cave systems that seemed very mature. This observation alone might indicate that the major cave pattern observed today was pre-glacial, becoming established in the Pliocene. The U/Th speleothem data indicate that the caves have, at the very least, survived multiple glaciations (Chap. 3; Lauritzen and Mylroie 2000).

The topography of the Cobleskill Plateau, when viewed at the 7.5-min quad map scale, has clearly been glaciated (Fig. 5.10). Numerous drumlins with a general northeast to southwest orientation are apparent at map scale, and in the field, glacial erratics, till, striations, and other indicators of glaciation abound. Boulders and cobbles in the till and in stream beds show provenances consistent with transport from the Adirondack Mountains to the north, across the Mohawk River. In local areas, the drainage is deranged, and the connection of surface flow to the cave systems below seems at times opportunistic. The role of glaciation on cave function in a new hydrology was the big question for researchers examining the cave systems of the Helderberg Plateau. That work was most productive in areas where large cave systems allowed patterns to emerge that were less obvious in smaller caves. For this reason the Schoharie County region, with the Cobleskill Plateau and Barton Hill cave development setting, was the most useful.

There is not much evidence in the Helderberg Plateau of extensive glacial quarrying of the rock. There are no U-Shaped valleys, and smaller features such as roches mountonées are rare. Striations are common, preserved on limestone surfaces that are covered by tills rich in carbonate fragments, which retards dissolution at the till/limestone contact (Fig. 5.11). It has been suggested that glacial stripping of weaker clastic rocks from the limestones had occurred (Mylroie 1977), creating new subaerial limestone exposures where none had previously existed. Minor glacial quarrying of limestone outcrops created a classic Yorkshire *clint and grike* limestone pavement in many places (Fig. 5.12). These limestone pavements initially were probably a typical pinnacle and cutter system as found throughout the interior lowland plateau limestones of the United States and elsewhere (Fig. 5.12a). Subsequent glacial quarrying (Fig. 5.12b–d) removed the pinnacles leaving flat bedrock behind as the clints, with the lower portion of the cutters remaining as the grikes (dissolutionally enlarged joints). There is a question regarding the rate of limestone pavement development. Given that glacial periods are 5–10 times longer than interglacials over the last 800 ka, is the interglacial time window long enough for cutter and pinnacle topography to develop prior to the next ice advance? If not, then the limestone pavements are likely inherited from interglacial to interglacial, modified to some extent by dissolutional denudation before the next ice cover event grinds them flat again, and the current features represent the integration of all those cyclic events.

The major alteration of the cave systems in the Helderberg Plateau was as a result of derangement of the surface drainage, both on a local scale and on a regional scale. This derangement affected both inputs (insurgences) and outputs (resurgences) of the cave systems. In the Barton Hill area, Fox Creek has incised well below the level of the Helderberg Group

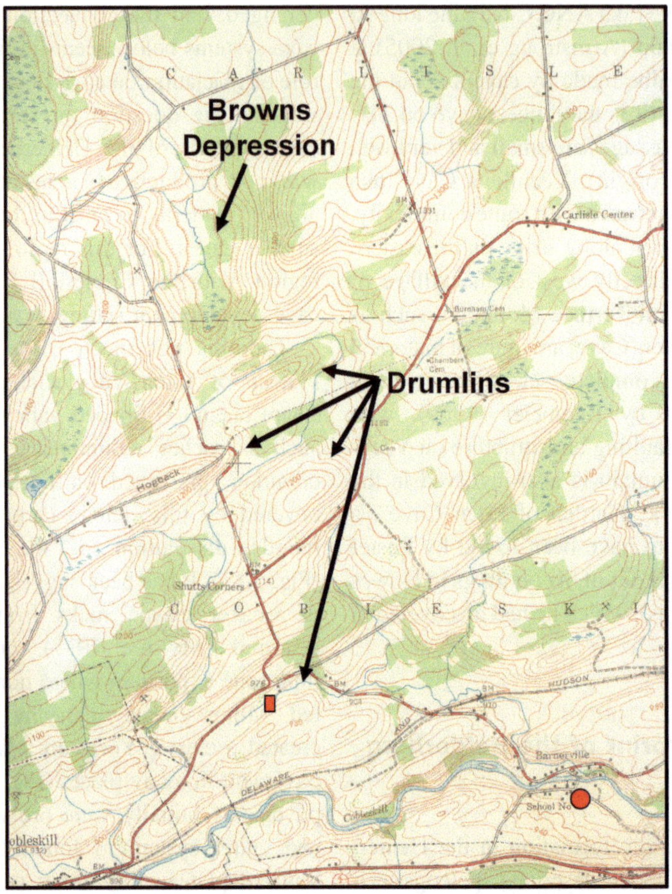

Fig. 5.10 Portion of the USGS 7.5 min Cobleskill quadrangle map displaying part of the Cobleskill Plateau showing major topographic features. Numerous drumlins are visible, striking northeast to southwest. Many topographic closed depressions are present, the largest of which is Browns Depression (note the stream is shown sinking in the center of the depression). *Square* marks the location of Doc Shauls Spring (Fig. 5.2); *circle* the location of the Sinks of the Cobleskill. Compare with Fig. 5.7

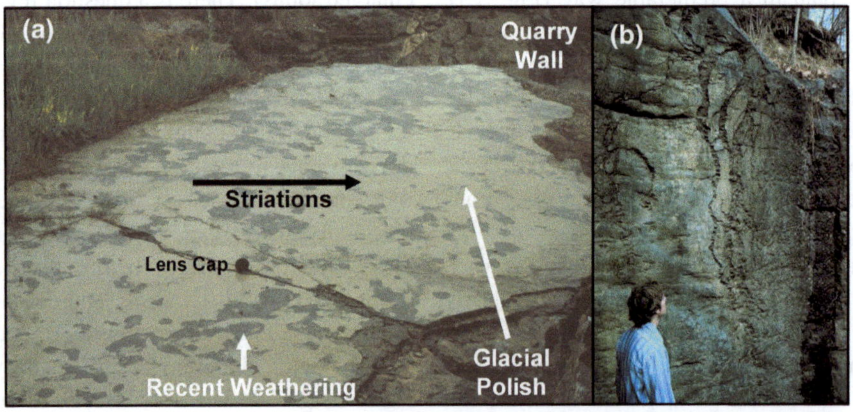

Fig. 5.11 Glacial polish and subsequent dissolutional removal, Schoharie County, New York. **a** Glacial polish, showing striations, developed on the Coeymans Limestone on the Cobleskill Plateau, looking south. The covering glacial till was removed prior to blasting in a quarry (background, right); the till is rich in carbonate clasts which protected the limestone surface from dissolution. The *darker gray patches* show where the glacial polish has been removed by recent dissolution since uncovering about 10 years earlier; the *lighter gray areas* show striations running *left* (northeast) to *right* (southwest). **b** Cliff in the Coeymans Limestone exposed by excavation, Schoharie Caverns (see Fig. 5.13). Water trickling down the newly exposed, glacially polished cliff face (~40 years old at the time of the photo) has dissolved karren channels into the rock

Fig. 5.12 Surface karren and glaciation. **a** Pinnacle and cutter development in the subsoil environment, Miocene Aquada Limestone in a road cut in Puerto Rico. **b** Clint and grike development on the massive Devonian Coeymans Limestone, Barton Hill, New York. **c** Clint and grike development in Cambro-Ordovician marbles, northern Norway. **d** Clint and grike development in the Lower Carboniferous (Mississippian) Great Scar Limestone, Yorkshire, United Kingdom. Clints and grikes are essentially glacially planed pinnacles and cutters, respectively

limestones, and the cave resurgences are perched on Ordovician clastics that underlie the limestones as obvious springs. Even so, these springs commonly occupy reentrants in the cliff face called *spring alcoves* (Palmer 2007), which collected glacial till. The original outlet for Schoharie Caverns (Fig. 5.6) was masked by glacial till, and entry to the cave was difficult until after that material was excavated in the 20th century (Fig. 5.13). Further east, the limestones' southern boundary is mantled with till, and resurgences are pervasively blocked with glacial sediments and overflow routes have developed to release the cave water (e.g. Thunder Hole and Skull Cave). On the Cobleskill Plateau side, the downstream portion of Cobleskill Creek near its junction with Schoharie Creek has cut below the limestones, and a few cave systems in this area drain freely from the valley wall. Upstream to the west, drumlins and other glacial sediments were deposited on the approximate thalweg of pre-glacial Cobleskill Creek (Fig. 5.10), and the large cave systems there, McFails Cave and Barrack Zourie Cave, drain by way of overflow routes (Fig. 5.7).

The glacial sediment blockage of resurgences that were either graded to a pre-glacial stream, or were perched on underlying clastics, had a major impact on the cave systems that feed them. To the east of Barton Hill, the Skull Cave system experienced significant backflooding which produced a maze of enlarged joints up to 20 m (65 ft) high, and a series of large flood overflow tubes that redistributed water to newer and higher resurgences (Fig. 5.8). On the west side of the Cobleskill Plateau, the water from both McFails Cave and Barrack Zourie Cave resurges at Doc Shauls Spring, the major karst spring in the area (Figs. 5.2c and 5.7). The spring is a vertical pathway piped up through glacial sediments for ~30 m (100 ft) from the original cave spring in the buried valley wall below (Fig. 5.14). Inside McFails Cave, flood over-flow passages and enlarged joints are less prominent than in Skull Cave far to the east, but in Barrack Zourie Cave, enlarged joints are present and flood over-flow passages common.

Glacial sediment in the original valley of Cobleskill Creek has pushed the stream south, up and onto a structural bench in the Helderberg Group Limestones, developed on the massive Coeymans Limestone on the south valley wall. With the bedrock dip 2° SSW, the Coeymans Limestone has "held" Cobleskill Creek in this position, not allowing the stream to re-enter its

Fig. 5.13 Schoharie Caverns, Schoharie County, New York. **a** Photograph of the entrance to Schoharie Caverns in Barton Hill, taken to celebrate the show cave's grand (but short-lived) opening on July 26, 1958. Miss America, Marilyn Van Derber, leads the procession. The cave entrance in the background is formed at the Coeymans-Manlius limestone contact. The capping massive Coeymans Limestone displays glacial scour, and the glacial till blocking the original entrance is clearly visible at the cliff face and along the walkway (*Source* Northeastern Caver, v. 44, no. 1, cover photograph; photography reproduced from the James L. Gage files, possibly taken by Russ Gurnee). **b** Schoharie Caverns entrance photographed in 1976, showing slumping of glacial till (a culvert laid in 1958 carries water under the debris). The glacial polish of the outcrop is well displayed; ice moved from the northeast (*right*) to the southwest (*left*) see also Fig. 5.11b. The original entrance was a slide down the outcrop face; a tapoff passage took (and still takes) water around the glacial blockage to Tufa Spring, down valley (Chap. 1, Fig. 1.4a)

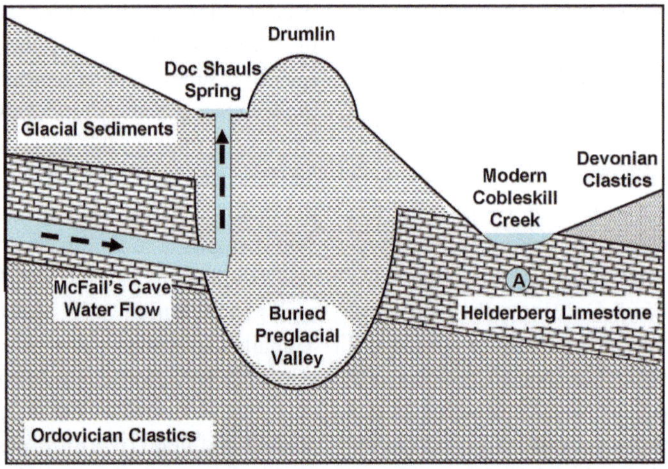

Fig. 5.14 Diagrammatic representation of the flow system at Doc Shauls Spring, looking east. The original resurgence on the north valley wall of Cobleskill Creek has been blocked by a large drumlin along the valley axis (Fig. 5.10), forcing the water to rise approximately 30 m (100 ft) upward through glacial sediment for release. The modern Cobleskill Creek has been displaced up onto the bedrock bench of the south valley wall, and has been "pinned" there by the SSW dip of the rock. Cave development has initiated in the south valley wall (point *A* on the figure) to the extent that all low-flow water bypasses the surface channel for 1.6 km (1 mile)

original mid-valley position, and therefore preventing the stream from clearing its original stream course of glacial sediment (Fig. 5.14). In response to this situation, Cobleskill Creek, at low water conditions, sinks into the Coeymans Limestone where it is first encountered, and the water resurges 1.6 km ($\sim$1 mile) to the east from the underlying Manlius Limestone (Fig. 5.7).

The derangement of surface drainage has many implications. The largest and most significant example is the paleo-valley that stretched from near the northern edge of the Helderberg Plateau on the west side of the Cobleskill Plateau southward to a junction with Cobleskill Creek (Figs. 5.5d and 5.7). This valley was obstructed by a variety of glacial sediments (Palmer 1976), including drumlins that cross from one valley wall to the other (Fig. 5.10). A number of closed depressions were created, but unlike other glacially obstructed areas in New York (e.g. the Finger Lakes), these depressions did not become lakes. Limestone outcrops in the walls of these depressions took the surface flow underground into pre-existing cave systems. While initially thought to be large dissolution sinkholes, these depressions were eventually explained as blocked valley segments (Kastning 1975), which removed a question about how much dissolution would have been necessary to create such a large sinkhole. Browns Depression, along the axis of the pre-glacial valley, is such a feature, 1300 m (4300 ft) long, north to south, and 450 m (1500 ft) east to west in dimension, and 20 m (60 ft) deep (Fig. 5.10). The topographic map shows local drainage sinking at the center of the depression, but the water actually flows to the west wall of the depression and sinks into small ledges of exposed Coeymans Limestone, which represent upper portions of the pre-glacial valley wall.

Analysis of water well data, accompanied by gravity, seismic refraction, and resistivity surveys, demonstrated the existence of the pre-glacial valley and determined its general north-south orientation (Palmer 1976; Mylroie 1977). The water in Browns Depression was originally thought (Baker 1973, 1976) to drain to the southeast by way of the Northwest Passage in McFails Cave, a segment of the major trunk master cave that continued on to Howe Caverns (Figs. 5.2, 5.4, and 5.7). Dye tracing from Browns Depression had demonstrated a link to Doc Shauls Spring, the known resurgence for McFails Cave, so the interpretation was likely (Baker 1973, 1976). However, the sink point of the water in Browns Depression into the west, and not east, side of the depression, indicated a different interpretation might be possible. The original dye trace exercise did not place dye detectors in the Northwest Passage of McFails Cave; later dye traces demonstrated that the Northwest Passage water came from Sellecks Cave (Fig. 5.7) and other caves to the north of the end of the Northwest Passage (Mylroie 1977). These caves are all located on the east bank of the pre-glacial valley. Based on this dye test, Mylroie (1977) predicted that an additional cave must lie under the west wall of Browns Depression. This hypothetical cave, Barrack Zourie Cave, was subsequently discovered and found to contain 5 km (3 miles) of passage that dye tracing revealed drained south to Doc Shauls Spring (Dumont 1995). Several other depressions to the north, also on the west side of the pre-glacial valley (e.g. Cave Mistake, Fig. 5.7), also dye tested to Doc Shauls Spring (Dumont 1995).

These results created an interesting problem. It was unclear if Doc Shauls Spring was on the east or west side of the pre-glacial valley, but in any case, one of the two large cave systems of Barrack Zourie Cave and McFails Cave must drain through the glacial sediments infilling the pre-glacial valley to reach Doc Shauls Spring. The many backflooding features in Barrack Zourie Cave, and the relative lack of them in McFails Cave, suggests that Doc Shauls Spring is on the east side of the pre-glacial valley, and that the Barrack Zourie Cave water works its way through coarse glacial sediments infilling the pre-glacial valley to reach the Doc Shauls Spring release point, as shown in Fig. 5.7. The travel path from McFails Cave to the Doc Shauls Spring water exit is a short and direct route from the buried Cobleskill Creek valley wall directly upward 30 m (100 ft) to the surface (Fig. 5.14). This short, open route removes hydrological inefficiencies that would promote significant backflooding within the cave during major discharge events. The Barrack Zourie Cave water, on the other hand, must traverse the glacial sediments of the infilled, pre-glacial valley. Hydraulic inefficiencies there would promote extensive and major backflooding in Barrack Zourie Cave, which is what is observed in the cave (Dumont 1995).

As has been described, many cave resurgences (karst springs) were blocked or occluded by glacial sediment. Sometimes, as with Schoharie Caverns, that blockage was local, and in some cases, as for McFails Cave and Barrack Zourie Cave, the blockage was extensive. The impedance of flow inside the caves has in some cases created a dissolutional overprint of the original cave stream pattern. As previously noted, McFails Cave shows (relatively) minor backflooding modification, Barrack Zourie Cave much more so, but the most spectacular example is Skull Cave (Fig. 5.9). The current map of Skull Cave shows some sinuous passages that relate to the original dendritic flow pattern along strike and then down-dip towards Fox Creek. The emplacement of major amounts of glacial till blocked that original spring discharge point, and water escape could only be accomplished by water rising up so as to flow out of the limestones above the till blockage. At low flow conditions, water can escape to some extent through a few lower release points, but during major discharge events the cave fills with aggressive floodwater under high hydraulic heads. The result has been development of a large floodwater maze on top of the dendritic stream system, with large tubes that transfer water laterally along strike (Palmer 1975, 2007). Unloading due to deglaciation has been hypothesized to allow joints in the massive Coeymans Limestone to expand, providing a preferential route for groundwater to follow during flood conditions (Kastning 1975; Mylroie 1977); this idea of joint gapping has been utilized to explain maze caves in New York and in Norway (e.g. Cooper and Mylroie 2014 and references therein). That pattern is well revealed by the Skull Cave map, on which the original dendritic stream pattern flowing south can still be discerned, with long, narrow passages trending north-northeast representing the backflood-enlarged joints (Fig. 5.9). In the south of the map, the larger passages trending east to west represent large oval flood overflow tubes. The result is a cave with significant and extensive passage diversity. Glaciation, by creating joint gapping and backflooding, has acted to increase cave size and cave complexity, counter to the early thoughts regarding glaciation and cave destruction preceding the 1970s. Knox Cave also drains to the Skull Cave resurgences, but is far up-dip and while it displays significant joint control, the cave does not experience the flooding seen in Skull Cave due to its up-dip position. West of Skull Cave, the 3 km long Thunder Hole resurges from the Coeymans Limestone uphill from Fox Creek, as glacial sediments block the original resurgence. The cave shows long fissure passages and copious muds as a result of backflooding (Figs. 5.8 and 5.15).

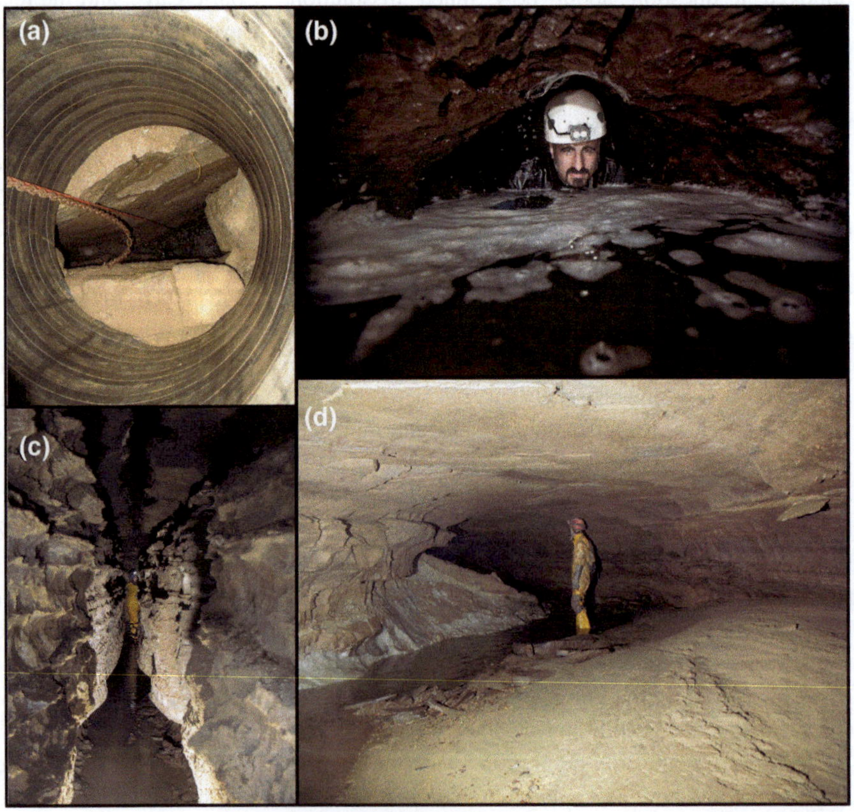

Fig. 5.15 Access to Thunder Hole, Schoharie County, New York. **a** Culvert leading down to excavated section of the entrance series. **b** Small, immature passage conducting deranged surface drainage into the cave. **c** Joint-controlled passage in the entrance series. **d** Main cave passage (Photos courtesy of M. Chu)

Where the blockage of springs has been very local, as at Schoharie Caverns on Barton Hill, many caves have reacted by creating new flow routes that bypass that obstruction. These bypasses have been called *tapoff passages* (Mylroie 1977, 1984a). The tapoff passage is not unique to glaciated regions, but the concept was developed there to assist in understanding how cave systems reacted and behaved after glaciation. In any karst area, cave entrances, such as springs, can be unstable as the interface between the surface and subsurface environments can promote breakdown, sediment deposition, and slumping that create blockages and hydraulic inefficiencies that require bypassing. The tapoff passage provides that bypass route. In nonglaciated areas, it is common to enter the cave by climbing over a debris pile, only to see the stream inside flowing out of the cave by entering a small passage that bypasses the debris pile and flows to a nearby spring. In glaciated regions, the debris pile can be glacial till that entirely conceals the cave entrance. In those cases, the tapoff passage becomes the only access point for human exploration. The type example is Single X Cave, on the south escarpment of Barton Hill (Fig. 5.6), where 275 m (900 ft) of small, wet stream passage leads from the cave spring into a large, pre-glacial cave system containing one of the largest rooms and passages in the northeastern United States. As that passage trends south to the escarpment above Fox Creek, it is completely blocked by unsorted glacial sediments. On the escarpment, this point can be seen as a break in the cliff line infilled by glacial till. Schoharie Caverns (Figs. 5.6 and 5.13) has a similar tapoff passage (not fully passable by humans to the spring).

Other caves have been discovered by recognizing that flood overflow springs should exist when glaciation masks original discharge points. Caboose Cave (Fig. 5.6) was discovered by applying this concept to a dry streambed that went uphill to a Coeymans Limestone outcrop. While the Coeymans Limestone is an excellent caprock, and many cave passages are roofed by it, most cave passage is actually in the thin-bedded Manlius Limestone beneath the Coeymans. In other words, cave resurgences in the Coeymans Limestone in the flat-lying Helderberg Plateau were unexpected. But at Caboose Cave, backflooding had produced an overflow spring at the Coeymans/Manlius contact, and the cave stream at low flow conditions is 10 m (30 ft) below the entrance inside the cave. At high flow, the water exits the cave by way of the overflow spring.

The input of water into the pre-glacial cave systems was also highly modified by the derangement of surface drainage patterns. As a first impression, a chicken vs. egg situation could be considered to exist. Were the present insurgences the result of where the limestone had been left clear of glacial till, such that water could enter the limestone and then join the cave system, or did the presence of a pre-existing cave system control subglacial water flow and sediment deposition, such that upon ice retreat those entry points were open to receive surface flow again? Field observations seem to indicate that three options exist. Blocked and occluded former entry points for surface water are found in many caves, such as Northwest Dome in the Northwest Passage of McFails Cave, which when bolt climbed, was found to be completely blocked by glacial till. Other water entry points, such as the McFails Hole entrance to McFails Cave, are large and directly overlie the main cave stream passage, suggesting it has been there a long time and was preglacial. Other cave access points seem immature, being diminutive in size, and connecting by small passages to the current cave system, such as the Barn Entrance to Barrack Zourie Cave and the mined entrance following a small surface water input to gain access to Thunder Hole (Fig. 5.15), suggesting both inputs are post-glacial in origin. It seems that the surface water does what would be expected: it no longer utilizes entrances that are completely blocked with glacial sediments, but it does use pre-existing entrances that remained open or were able to be flushed of occluding sediment, or the water exploited the well-developed and glacially reactivated joint system in these limestones to forge new post-glacial flow paths into existing caves.

While post-glacial inputs to existing pre-glacial caves seem to exist, have entirely new post-glacial caves developed? Mylroie and Carew (1987) proposed that the time of origin of a cave in the Helderbergs could be determined by the degree to which the cave, its tributaries and inputs, and its output point, were in agreement with the modern glacially deranged surface drainage (Chap. 3). Palmer (1972) had suggested that Onesquethaw Cave might be post-glacial. U/Th dating of speleothems from that cave found none over $\sim$5 ka in age (Lauritzen and Mylroie 2000); not proof but certainly supportive of A. Palmer's speculation. Weremeichik and Mylroie (2014) have postulated that Westfall Spring Cave on the south side of Barton Hill is postglacial as it is small in dimensions, in agreement with current drainage patterns, and does not contain lacustrine glacial sediments found in neighboring caves. Cooper and Mylroie (2014) have proposed that Barbers Cave, a small maze cave on the east side of Barton Hill (Fig. 5.6), is post-glacial based on agreement with current drainage patterns, passage size, discharge determinations, and passage formation rates. There is no reason why post-glacial caves could not exist, especially those that form at high rates such as floodwater maze caves. The 13 ka time since the complete deglaciation of New York is theoretically long enough for caves to reach humanly traversable size, based on first-principle dissolution kinetic analyses (e.g. Palmer 2007). The interesting turn-about in this discussion is that the original consensus prior to the 1970s was that caves could not survive glaciation (Mylroie and Mylroie 2004; Chap. 3). The research done in the 1970s clearly demonstrated that the large caves of the Helderberg Plateau were pre-glacial in origin, in agreement with data from Yorkshire, Norway and elsewhere. The pendulum then swung to the idea that there had not been time enough for caves to form post-glacially. The recent research described above seems to indicate that both pre-glacial and post-glacial caves exist in the Helderberg Plateau, and by extension, throughout New York and New England (Chap. 8).

5.2.2 Glaciation and Cave Sediments

The application of large volumes of glacial sediment to the land surface overlying pre-glacial caves in the Helderberg Plateau should have resulted in transport of some of those sediments into the cave systems below. Blocked vertical shafts, such as Northwest Dome in McFails Cave, show that till could be extruded directly into the caves. The great abundance and variety of cobbles and gravels in the caves indicates that till was washed into the caves. Some of those cobbles and gravels have provenance from the Adirondack Mountains north of the Mohawk River, and as a result could not be autochthonous to the Helderberg Plateau. Therefore, glacial sediment transport into existing caves is demonstrated (no caver who has ever crawled over these gravels and cobbles in a Helderberg cave would ever disagree; Figs. 5.9c and 5.16a). The backflooding experienced by many caves results at times in stagnant flow conditions in the caves, with the subsequent deposition of fine grained sediments, producing ubiquitous dark-brown mud in most cave passages that are not active stream channels (Figs. 5.16b and 5.17d). Many passages inside the caves of the Helderberg Plateau are blocked with sediments of glacial origin. These blockages introduce further hydraulic inefficiencies into the cave flow systems,

Fig. 5.16 Cave sediments in Helderberg Plateau caves. **a** Coarse-grained sediments in Caboose Cave (Fig. 5.6); some cobbles have provenance from the Adirondack Mountains; (photo courtesy A. Palmer). **b** Cave explorers washing off fine-grained muds that cover many passage walls in Barrack Zourie Cave (Fig. 5.7) as a result of backflooding and stagnation of water flow (Photo courtesy of K. Dumont)

Fig. 5.17 Cave sediments in Thunder Hole, Schoharie County, New York. **a** Gravels with provenances both local and from the Adirondack Mountains; note the thin-bedded Manlius limestone forming the cave walls. **b** Small side passage discharging sulfur water into the cave. **c** Light-colored, thin-bedded glacio-lacustrine sediments overlain by coarse gravels; these deposits are the most easterly known from the footprint of Glacial Lake Schoharie. **d** Downstream in the cave, where backflooding deposits mud throughout the passage (Photos courtesy of Mike Chu)

resulting in local backflooding within the cave independent of the flow situation at the cave resurgence.

The question becomes, when were these sediments introduced into the caves (Mylroie 1977)? Was it during ice advance, when end moraine material would have been bulldozed over the caves, during ice cover when ice motion would bring basal tills over cave entrances, during ice retreat, when abundant melt water would mobilize moraines, kames and till sheets, or did large-scale sediment introduction wait until post-glacial times, when "normal" flow resumed through the cave passages (the "car wash" model of Lauritzen and Skoglund 2013)? It is clear that the last case is operating today, as re-worked glacial sediments can be seen entering caves during modern flood conditions. It is more difficult to parse out the glacial events. Mylroie (1984b) attempted to demonstrate that very fine-grained deposits found in many Cobleskill Plateau and Barton Hill caves were the result of deposition during ice cover. These sediments, very white or light tan in appearance, had a high $CaCO_3$ content, and were very fine grained, approaching clay size. The deposits are formed of many very thin layers, which give them a varve-like appearance, however lacking the couplet grain size variation found in true varves (Figs. 5.17c and 5.18c). Given the high surface to volume ratio of such tiny carbonate grains, it was unlikely that they could long survive dissolution in a turbulent cave stream environment. In the modern cave environment, such deposits are cohesive and resistant to mechanical erosion and transport. Along stream channels, these deposits are cohesive enough, and contain enough soluble components, to display ablation scalloping (Dumont 1995). The hypothesis (Mylroie 1984b) was that the carbonate material was glacial rock flour locally produced from the Helderberg Plateau, and that it had settled out, and been preserved, by the stagnant conditions which would have existed in these caves when ice was more than a kilometer thick above the cave, and ice extended south to the Pennsylvania border. The problem with this interpretation is that such carbonate rock flour deposits were only found in the Cobleskill Plateau and Barton Hill areas (e.g. Rubin 1991a), not elsewhere in New York or New England, nor in Yorkshire. A large-scale follow up project (Weremeichik and Mylroie 2014)

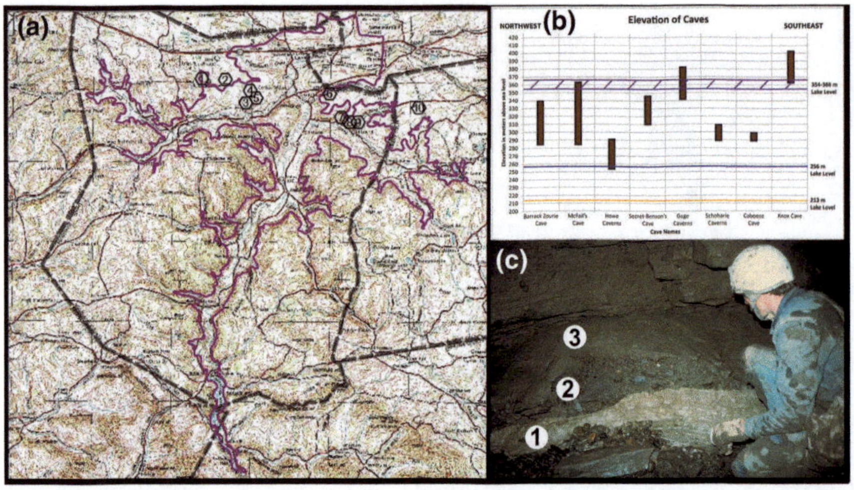

Fig. 5.18 Glacial Lake Schoharie cave sediments. **a** Topographic map with the outline of Schoharie County (*bold dashed line*) with the footprint of Glacial Lake Schoharie at the 354–366 m level outlined in *purple*. Selected caves on the Cobleskill Plateau and Barton Hill shown as numbers in hexagons, in the same order as shown in **b**, except that **b** shows Secret-Bensons Cave as a single cave, and does not show Westfall Spring Cave (post-glacial cave between Caboose Cave and Schoharie Caverns in a; see also Figs. 5.6 and 5.7). Knox Cave is in adjacent Albany County to the east, outside the lake footprint. **b** Elevation of the caves used in the sediment study. Although the entrance to McFails Cave shows as outside the footprint of the lake in a, it extends well below lake level. Howe Caverns, which has the thickest glaciolacustrine deposits, is at the lowest elevation and would have been under the lake longer than any other cave pictured. **c** Typical sediment sequence from Caboose Cave. *1* glaciolacustrine clays; *2* coarse re-worked glacial sediments from ice withdraw and post-glacial (and post-lake) time; *3* dark, organic-rich muds perhaps related to European settlement, clear-cutting, and agriculture (Diagrams **a** and **b** from Weremeichik and Mylroie 2014)

provided an alternative, but closely related answer, that the deposits were lacustrine, deposited under Glacial Lake Schoharie at the end of the Wisconsin glaciation. By plotting lake levels and cave elevations (Fig. 5.18), it was shown that not only were all rock flour sites within the lake's footprint, but that the thickest deposits were in those caves that would have been under the lake level the longest, as lake levels fluctuated (e.g. Howe Caverns). The study also demonstrated, as noted earlier, that Westfalls Spring Cave, well within the lake footprint, lacked such deposits, evidence that the cave was postglacial. These results explained the lack of such rock flour deposits in caves elsewhere in the region, in New York, in the northeastern United States, and in Yorkshire, England.

5.3 Cave Exploration and Discovery in the Helderberg Plateau

As noted earlier, understanding the general, pre-glacial model of cave development in the Helderberg Plateau allowed prediction of where cave passages should be when the surface landscape and drainage had been subsequently deranged by glaciation. The discovery of Barrack Zourie Cave was predicted by the overall cave model (Fig. 5.5d). The discovery of Caboose Cave and Single X Cave was the result of applying lessons learned from how glaciation affects cave resurgences. None of these discoveries were easy, as in many cases glacial sediment had to be excavated to gain entry, as at Caboose Cave and Single X Cave. At Barrack Zourie Cave, not only was glacial sediment excavated from a water input point, but the glacially gapped bedrock joint had itself to be enlarged to gain entry to that cave system (Hopkins 1992). Once in the cave system, a second entrance was located from the inside, and the excavation was relatively easy compared to the initial entrance effort; this second entrance was not obvious from outside surface inspection. A similar approach of enlarging an immature, post-glacial water input (Fig. 5.8) allowed access to Thunder Hole (Armstrong et al. 2008). Skull Cave is an example of in-cave discovery in the 1970s as a result of understanding how the backflood mazes and overflow routes functioned in the post-glacial cave hydrology (Fig. 5.9).

Cave exploration in the Helderberg Plateau, as with the rest of New York and New England is arduous. The low mean-annual temperature (48 °F or 9 °C) means exposure is a danger; this danger is amplified by the abundance of water in the caves, and the restricted size of some of the passages. Wetsuits are often essential. The size restriction results from both immature, postglacial passages, and from partial occlusion of large passages with glacial sediments. The difficulty of negotiating small passages floored with gravels and cobbles was previously mentioned, made worse if one is in a constricting wetsuit. Many caves are entered by way of vertical shafts, which adds an additional hardship, as vertical caving gear must be utilized. One death at the Acks Shack entrance to McFails Cave was a combination of failure of a vertical technique combined with exposure (Stone 2014), a similar death occurred west of the Cobleskill Plateau at Schroeders Pants Cave (Speece 2006; see also Chap. 7).

Most early cave exploration in the Helderberg Plateau was done by simply entering cave entrances that were open and humanly passable. Serious attempts to gain cave access by digging out entrance obstructions did not begin to occur in a routine way until the late 1960s and early 1970s, when several major discoveries were made (e.g. Single X Cave, Caboose Cave). The original entrance to McFails Cave collapsed in 1963, and access was regained through the Acks Shack entrance by digging out sediment from a small, immature passage that lead from an enlarged surface joint down to the cave level and from there to the cave stream (Stone 2014). Digging to gain cave access has continued, with Barrack Zourie Cave in the 1990s and Thunder Hole in the 21st century being major accomplishments. Many of these digs were long, sustained efforts, and would have been unlikely to have occurred if the cave area had not been close to major population centers that allowed access to manpower and continued interest in the caving community. Many of the exploration advances in the last few decades have been assisted by tracing the hydrology of the karst flow paths, such as Mylroie's tests prior to the Barrack Zourie discovery (Mylroie 1977), or Woodell's work prior to the opening of Thunder Hole (Woodell 2004). Tracing has moved beyond dyes to radionuclide, stable isotope and dissolved solids analyses (e.g. Siemion et al. 2005).

5.4 Summary

The Helderberg Plateau contains mature, pre-glacial cave systems that were originally typical in plan and pattern to those found in Paleozoic limestone plateaus elsewhere in North America. These cave systems were repeatedly glaciated during the Quaternary. Despite repeated glaciation, these subsurface flow paths were able to re-establish themselves once ice cover retreated. The caves were modified primarily by backflooding, caused by glacial sediment that occluded springs and resurgences. This backflooding in many caves created flood water mazes and flood overflow routes that added to cave complexity. The effects of backflooding were enhanced by gapping of the major joint pattern due to isostatic rebound. Post-glacial caves have also developed, small in dimension and closely coupled to the current surface drainage regime.

The upland plateau surface drainage was significantly deranged, and older insurgences became senescent, while new ones developed. The Cobleskill Creek was diverted up onto a limestone bench, and promptly began making a new cave system in that bench. Glacial sediments were introduced, and continue to be introduced, into the caves. These sediments created hydraulic inefficiencies within the caves, leading to additional backflooding conditions. Fine-grained rock-flour deposits once thought to be sub-ice in origin are now recognized to be lacustrine deposits from a glacial lake.

Caves everywhere have long been known to be repositories of past geologic history, both by way of their flow patterns, and by their mineral and sediment contents. The caves of the Helderberg Plateau show all these features, plus the additional information provided by their glacial overprint. They not only carry a record of glaciation, but lessons in how pre-existing karst flow paths can reassert themselves despite repeated large-scale rearrangement of water inputs and outputs.

References

Armstrong J, Carley C, Hopkins T, Perlman S, Siemion J (2005) Another breakthrough at ANC Cave. Northeastern Caver 36 (4):125–126

Armstrong J, Chu M, Hopkins T, Kunz J, Millet S, Palmer A, Perlman S, Robinson A, Siemion J, Whittemore M (2008) Thunder hole reaches a mile. Northeastern Caver 39(3):84–94

Baker VR (1973) Geomorphology and hydrology of karst drainage basins and cave channel networks in east central New York. Water Resour Res 9:695–706

Baker VR (1976) Hydrogeology of a cavernous limestone terrane and the hydrochemical mechanisms of its formation, Mohawk River Basin, New York. Empire State Geogram 12:2–65

Cook JH (1906) Limestone caverns of eastern New York. In: Clarke JM (ed) Third report of the Director of the Science Division. New York State Museum, Albany, pp 32–51

Cooper MP, Mylroie JE (2014) Post-glacial speleogenesis: verification of a hypothetical model, and the origins of maze caves in glaciated terrains. Cave Karst Sci 41(2):84–95

Dumont KA (1995) Karst hydrology and geomorphology of Barrack Zourie cave system, Schoharie County. Bulletin V of the New York Cave Survey, New York

Egemeier SJ (1969) Origin of caves in eastern New York as related to unconfined groundwater flow. Natl Speleol Soc Bull 31:97–111

Engel TD (2009) New York State. In: Palmer AN, Palmer MV (eds) Caves and karst of the USA. National Speleological Society, Huntsville, Alabama, pp 27–32

Halliday WR (1976) Depths of the Earth. Harper and Row, New York

Hopkins T (1992) The final dig. Northeastern Caver 23(3):77–80

Kastning EH (1975) Cavern development in the Helderberg Plateau, East-Central New York. Bulletin I of the New York Cave Survey, New York

Lauritzen S-E, Mylroie JE (2000) Results of a speleothem U/Th dating reconnaissance from the Helderberg Plateau, New York. J Cave Karst Stud 62(1):20–26

Lauritzen S-E, Skoglund RØ (2013) Glacier ice-contact speleogenesis in marble stripe karst. In: Schroder JF (ed), Treatise on Geomorphology 6:363–396

Mylroie JE (1977) Speleogenesis and karst geomorphology of the Helderberg Plateau, Schoharie County. Bulletin II of the New York Cave Survey, New York

Mylroie JE (1984a) Hydrologic classification of karst landforms. In: LaFleur R (ed) Groundwater as a geomorphic agent. Allen and Unwin, Boston

Mylroie JE (1984b) Pleistocene climatic variation and cave development. Nor Geogr Tidsskr 38:151–156

Mylroie JE, Carew J (1987) Field evidence for the minimum time for speleogenesis. Nat Speleol Soc Bull 49(2):67–72

Mylroie JE, Mylroie JR (2004) Glaciated Karst: how the Helderberg Plateau revised the geologic perception. Northeast Geol Environ Sci 26(1–2):82–92

Palmer AN (1962) Geologic of the Knox Cave system, Albany County, New York. Thesis, Williams College

Palmer AN (1972) Dynamics of a sinking stream system: Onesquethaw Cave, New York. Natl Speleol Soc Bull 34:89–110

Palmer AN (1975) The origin of maze caves. Natl Speleol Soc Bull 34(3):56–76

Palmer AN (2007) Cave Geology. Cave Books, Dayton

Palmer MV (1976) Ground-water flow patterns in limestone solution conduits. Thesis, State University of New York at Oneonta

Ridge JC (2004) The Quaternary glaciation of Western New England with correlations to surrounding areas. In: Ehlers J, Gibbard PL (eds) Quaternary glaciations—extent and chronology—Part II: North America. Developments in Quaternary sciences, vol 2. Elsevier, Amsterdam, pp 169–199

Rubin PA (1991a) Modification of preglacial caves by glacial meltwater invasion in east-central New York. In: Kastning EH, Kastning KM (eds) Appalachian Karst: Proceedings of the Appalachian Karst symposium, Radford Virginia

Rubin PA (1991b) Flow characteristics and scallop-forming hydraulics within the Mill Pond karst basin, East-Central New York. In: Kastning EH, Kastning KM (eds) Appalachian Karst: Proceedings of the Appalachian Karst Symposium, Radford Virginia

Siemion J, Palmer AN, Kraemer TF (2005) Use of quantitative dye racing and water chemistry to characterize karst springs in the Helderberg Plateau, east-central New York. In: Geological Society of America, Abstracts with Programs, vol 37(1), p 68

Speece JH (2006) James G. Mitchell and the Schroeders Pants Cave tragedy. Northeastern Caver 37(3): 92–96

Stone F (2014) The history of McFails Cave, part four: McFails purchase, the NSS owns its first cave. Northeastern Caver 45 (2):59–62

Waltham AC (1974) Limestone and caves of Northwest England. David and Charles, Newton Abbot

Weremeichik JM, Mylroie JE (2014) Glacial Lake Schoharie: an investigative study of glaciolacustrine lithofacies in caves, Helderberg Plateau, Central New York. J Cave Karst Stud 76 (2):127–138

Woodell PA (2004) Hydraulic interpretation of limestone solution conduits from dye tracing and flow analysis, Schoharie County, New York. Thesis, State University of New York at Oneonta

6 Caves and Karst of the Deformed Siluro-Devonian Carbonates, South Central New York

Abstract

The Siluro-Devonian Helderberg Group carbonates and the mid-Devonian Onondaga Limestone of the Hudson River Valley were highly deformed by the Acadian Orogeny, which limited subsequent outcrop extent and catchment. Glaciation overprinted the tectonic structure and obliterated, modified, and engendered cave development. Ice movement was north to south, parallel to the structural grain and the adjacent Hudson River, and differentiating the controls of cave development is difficult. Clarksville Cave and Onesquethaw Cave appear pre- and post-glacial, respectively, despite similar sizes in excess of 1 km (0.63 miles), and both show significant structural control. Farther south, the Joralemon Park caves display maze development initiated post-glacially, adjacent to pre-glacial cave remnants in a deranged drainage setting. Pompeys Cave is developed in a limestone member of the Rondout Formation, a member not present west and north in the main Helderberg Plateau. The cave is a single stream-capture passage of large size with a backflood maze overprint upstream of a major collapse area; despite the large size, the cave otherwise fits the criteria of being post-glacial in origin. Mystery (Surprise) Cave at 3 km (2 miles) is the longest cave in this region, and sits near the southern glacial boundary and appears pre-glacial in origin. Numerous other small caves are strewn along the Hudson River Valley, and all show an interesting interplay between structure, lithology and glaciation.

6.1 Introduction

The band of carbonates that trace the eastern edge of the Allegheny Plateau in south central New York contains the same geologic units as the Helderberg Plateau, though a more complete record of certain formations of the Helderberg Group is present (Chap. 2, Fig. 2.7). Unlike the equivalent strata in the Helderberg Plateau, the rocks in this band are highly folded, faulted and tilted (Fig. 6.1), as they approach the western margin of the metamorphic suite of eastern New York and New England. This suite of folded and faulted rocks is termed the Hudson Valley Fold-Thrust Belt, and is the result of the Acadian and Allegheny orogenies (Chap. 2, Isachsen et al. 2000). This deformation is in contrast with the Helderberg Plateau (Chap. 5), and the relatively non-deformed Cambro-Ordovician carbonate strata of the northern lowlands (Chap. 7), where there are few structures and the stratigraphy is layer-cake. Though this is the case for

M.P. Cooper and J.E. Mylroie, *Glaciation and Speleogenesis*,
Cave and Karst Systems of the World, DOI 10.1007/978-3-319-16534-9_6,

Fig. 6.1 Tectonics and glaciation in the Hudson River Valley. **a** Steeply dipping limestone beds of the Helderberg Group, Rt 23, Catskill, Greene County, New York. **b** Folded Helderberg Group limestones in the bed of Catskill Creek. **c** Boulder of Onondaga Limestone, moved by glaciers, southern Albany County. **d** Remnant bedrock knob of Helderberg Group limestone, showing fault displacement

much of the Helderberg Plateau, the eastern portion does contain some structures that influence Onesquethaw and Clarksville caves. For this reason, as well as its interesting glacial adjustments, Clarksville Cave and Onesquethaw Cave will be explored here.

This band of carbonates consists of the same Late Silurian and Early Devonian strata as the Helderberg Plateau, as well as the Middle Devonian Onondaga Formation. The formations of the Helderberg Group and Onondaga Formation here are less continuous in outcrop, although more continuous than the lenses of marble seen in the Adirondacks and New England discussed later (Chaps. 8 and 9). As these rocks are highly folded, faulted, and tilted, karst formation within them is not as extensive, and as such cannot form large, continuous cave systems such as those seen in the Helderberg Plateau like McFails Cave and Barrack Zourie Cave. The smaller caves, and the difficulty of interpreting the exact interplay of glaciation and karst when structure is thrown into the mix made this region less desirable to the glaciated karst workers, than the adjacent Helderberg Plateau (Chaps. 3 and 5). The structural deformation, glaciation, and karst development do however make for interesting, albeit smaller caves. In addition to Clarksville Cave and Onesquethaw Cave, two other caves in this area, Mystery (Surprise) Cave and Pompeys Cave, from the highly deformed part of the region, do make the list of the longest 25 caves in New York (Chap.1, Table 1.1).

Despite the added input of structure, several studies have been performed in this deformed band of carbonates. Clarksville Cave has been examined for supposed influences of meltwater on certain morphologies within the cave (Rubin 1991a, b). Onesquethaw Cave is controlled by faulting and folding, and is the focus of an early paper by Art Palmer (1972), who reconstructed the stages of its development, with particular attention paid to the effects of floodwater. The main study area of Cooper and Mylroie (2014), Joralemon Park, is located in this band. Additionally, site-specific work on piecing together the controls on speleogenesis of Pompeys Cave by undergraduates at SUNY New Paltz, under Dr. Alex Bartholomew, has taken place (Bowles and Bartholomew 2013; Bowles et al. 2014).

Similar to the dissertation by Mylroie (1977), the focus of this chapter will be on a smaller karst area. Here multiple small karst areas of this band will be explored, as the units in this region are non-continuous. Unlike Mylroie (1977), which spanned several topographic quadrangles, some of these karst areas only span small sections of an individual topographic map. Each of these smaller karst areas can represent a microcosm of glaciated karst, with a combination of glacially modified pre-glacial caves and post-glacial caves and surficial karst. A tour of this region begins with the caves of the eastern Helderberg Plateau, and wraps around towards the south with the caves and karst of Joralemon Park, and moves further south to Ulster County with Pompeys Cave.

6.2 Geologic Background

The area east and south of the Helderberg Plateau is part of the Hudson Valley Fold-Thrust Belt (HVFTB) (Marshak 1986). This thrust belt is a series of anticlinal ridges, and synclinal valleys, such as those seen in the Ridge and Valley province to the south in Pennsylvania, the Virginias, and other states. The Ridge and Valley deformation occurred over the Acadian and Allegheny orogenies, though to what ratio of impacts these had on the HVFTB is still being determined. Unlike the ridge and valley topography to the south, the HVFTB has been eroded over multiple glacial cycles, combined with the earlier non-glacial erosion following the Paleozoic. Though this glacial erosion occurred, there is little in the way of glacially incised valleys, as mentioned in Chap. 8, except for the Hudson River itself, which is considered by some to be a fjord (Isachsen et al. 2000). As this relative lack of glacial valleys is the case, particularly in the area east and south of the Helderberg Plateau, several streams such as Onesquethaw Creek maintain their general, pre-Wisconsinan routes (Dineen 1987). Glacial modifications do occur to the surficial drainage routes, such as the perching of tributaries on glacial tills (Palmer 1972).

The stratigraphy of the region is the same as the Helderberg Plateau to the west, with two of the caves discussed here being on the eastern margin of the Helderberg Plateau. The stratigraphy of this area has been described by Ruedemann (1930), as well as by Goldring (1943). Important Late Silurian-Early Devonian units are the Helderberg Group of limestones, and the underlying Rondout Formation. Early-to-Mid Devonian stratigraphy includes the Esopus Shale, the Schoharie Formation, and the Onondaga Formation. Above the Onondaga Formation is a series of clastics, with the basal Hamilton Group of shales. On the eastern portion of the Helderberg Plateau, where Clarksville and Onesquethaw caves are located the Onondaga Limestone is bounded between the Schoharie Formation (a clastic-rich carbonate unit) or the Esopus Shale, and the overlying Hamilton Group. A similar situation occurs to the southeast at the Joralemon Park karst area. Approaching southward, the Rondout Formation and Helderberg Group are found, as seen in the Pompeys Cave area. Formations within the Helderberg Group, and members within the Rondout Formation can act as dissolutional base level, or act as overlying, insoluble beds.

6.3 Eastern Helderberg Plateau—Clarksville Cave and Onesquethaw Cave

To the west the Helderberg Plateau has very little structure, and a shallow dip of 1° to 2° (Chap. 5). In the eastern area the plateau approaches the HVFTB, where folds and thrust faults begin to be seen. Onesquethaw Creek follows the structure of this eastern portion and routes through and near towns such as Clarksville, NY in Albany County. As noted above, this creek seems to closely track its original, pre-glacial flow route. Along this creek are several caves, the most spectacular and largest of these being Clarksville Cave and Onesquethaw Cave (Chap. 5, Fig. 5.1). Though along the tributaries of the same creek these caves are interpreted to be of different origin times, with Clarksville Cave having clear evidence of glacial modification (Kastning 1975; Rubin 1991a, b), and Onesquethaw Cave aligning with post-glacial deranged drainage, including a tributary perched on glacial till (Palmer 1972). These caves both have strong structural influences from the HVFTB, with both Clarksville Cave and Onesquethaw Cave being partially guided by faults. The close proximity of these caves, their control by structures, and their control by glaciation allow landscape evolution to be pieced together from the early seminal work on floodwater caves by Art Palmer (1972) on Onesquethaw Cave, and from a study by Rubin (1991a) detailing the evolution of Clarksville Cave, and in particular, hypothesized meltwater features within the cave. These caves are among the few within the HVFTB to receive attention in the literature. There is some disagreement with the findings of Rubin (1991a, b) and the interpretation presented here in this chapter.

Other caves do exist in the area, including relict, pre-glacial caves; as well as post-glacial caves that act as tributaries to the pre-glacial Clarksville Cave.

Of note, Onesquethaw and Clarksville caves have formed in the Onondaga Limestone, and are the longest known caves found within that unit in New York.

6.3.1 Development of Clarksville and Onesquethaw Caves

Clarksville Cave (Fig. 6.2) is located in the northwestern portion of the Hudson Valley Fold-Thrust Belt, and is the most northwestward cave with strong structural influences (Rubin 1991a). This cave has a length of ~4800 ft (~1460 m), with some phreatic passages being up to 15 ft (4.5 m) tall, and 40 ft (12 m) wide, making it one of the largest caves in the HVFTB. Historically this cave was two separate caves, Ward Cave, and Gregory Cave, which were connected; as a result it is also called Ward-Gregory Cave colloquially. The use of "Ward Section" and "Gregory Section" refers to the upstream and downstream portions of the cave, respectively (Fig. 6.3).

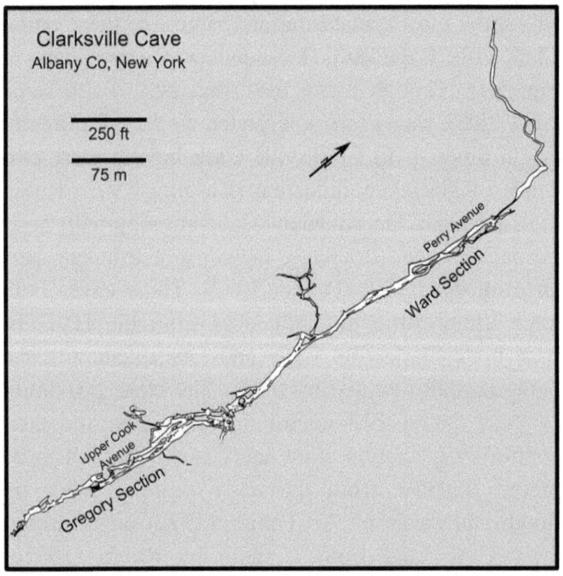

Fig. 6.2 Map of Clarksville Cave, Albany County, New York. This cave is the result of the connection of two caves: Gregory Cave, and Ward Cave. These two caves were connected through a breakdown pile. This breakdown pile creates a restriction, resulting in inefficiencies that, combined with glacial sediments, likely formed the backflooding features such as superimposed maze-work seen in the Gregory Section of the Cave. The plan view presented here shows the trend of the cave following a north-south oriented fault, though sumped passages upstream of Perry Avenue do not follow this trend. (Adapted from Nardacci 1991)

Clarksville Cave contains three definite phreatic levels following a north-south striking fault. The upper level is entirely contained within the Ward Section of the cave. The middle and lower levels of the cave are located within the Gregory Section of the cave. Even under the highest discharges only the lowest level floods to the ceiling. Despite this, floodwater features exist in the upper level passage within the Ward Section in the form of maze-work off of the main trunk passage. Other side passages contain vadose canyons and domes. Multiple phreatic levels offer support for this cave being pre-glacial, as the timeframe and base level changes since deglaciation are not enough to account for them.

Both Upper Cook Avenue and Perry Avenue (Fig. 6.2) contain sediments, with some cases of thick sediment packages indicating sizable passage before infilling (Rubin 1991a). These thick sediment packages on two phreatic levels supports a pre-glacial origin of the cave, as the time frame of the Holocene is not enough to establish large passages, infill them with sediment, and then subsequently remove some of the sediment. The sedimentation likely occurred over multiple glacial cycles, with large flood pulses such as meltwater flushing sediment out.

Rubin (1991a) offers that much of the speleogenesis of Clarksville Cave is due to glacial meltwaters; the degree of that meltwater impact on cave development is not quantified. Though flushing of sediment via glacial meltwater, and the establishment of floodwater maze-work due to backflooding are both important in the speleogenesis of this cave, these meltwaters are not likely to have entirely formed the upper levels. Speleogens such as canyons and domes, used to demonstrate meltwater genesis here, are not completely useful for determining the age of caves, as the heights and widths of such are not controlled simply by dissolution rates (Covington 2014 and references therein). The speleogenesis of Clarksville Cave appears mostly like the larger caves of the Helderberg Plateau to the west. Multiple phreatic levels were established over multiple glaciations due to base level change, with sediment infilling, flushing, and the establishment of floodwater mazes superimposed on a branchwork geometry. Glacial meltwaters have been offered as an important speleogenetic agent in other settings (Faulkner 2009; Quick 2010; Cooper and Mylroie 2014), and they were likely important to the development of Clarksville cave. The current

Fig. 6.3 Photographs of Clarksville Cave, Albany County, New York. **a** Main entrance, located in a collapse sinkhole. **b** Perry Avenue, the master cave passage in the Wards section of the cave; the ceiling is a chert bed. **c** Anastomotic development along the fault plane at the downstream end of the Wards section, where the collapse shown in **a** has caused diversionary passages to develop. Note the gravel is a mixture of authigenic angular chert fragments and rounded allogenic glacial gravels. **d** The Gregory section of the cave, looking up a phreatic tube formed on the fault plane, which is dipping towards the viewer. This cave receives a large amount of recreational caver visitation, as seen here. (Photos **a–c** courtesy of A. Palmer)

Holocene development of this cave mirrors that of the Helderberg systems, recapturing surface streams and adjusting to a new base level, with minimal substantial evolution. The segmentation of the cave into the Ward and Gregory sections is the result of a large collapse that created the middle entrance to the system (Fig. 6.3a). This collapse has helped create a floodwater diffusion maze in the cave (Fig. 6.3c).

Onesquethaw Cave (Fig. 6.4) on the other hand, shows substantial evolution through the Holocene. It is a long cave (1.6 km, 1 mile) for a post-glacial cave, with complex passages (Fig. 6.5). This cave may be entirely post-glacial in origin, as the initial passage is formed by a tributary perched on Wisconsinan till (Palmer 1972). The idea that determining if a cave is post-glacial by observing alignment with post-glacial, deranged drainage guided by glacial erosional and depositional features made one of its first appearances in the literature through Art Palmer's 1972 paper on Onesquethaw Cave. He broke the evolution of this cave into four stages, with a varying degree of structural and floodwater influence at each stage.

The first stage is the initial formation of the cave, with water being channelized by the perched tributary (Fig. 6.6). The portion of passage that accepted this water is now the entrance series of the cave, accepting the current stream at mid-to-high water (Fig. 6.5a, b). This initial development occurred along a shallowly dipping bedding plane. The dissolutional widening of this bedding plane pirated water from a surface stream, greatly influencing the further history of the cave. This pirating may have occurred in stages, indicated by two parallel passages along the same bedding plane (Fig. 6.4). With the capture of the surface stream, the second stage began forming most of the current flow path of the cave. During this stage the pirated waters

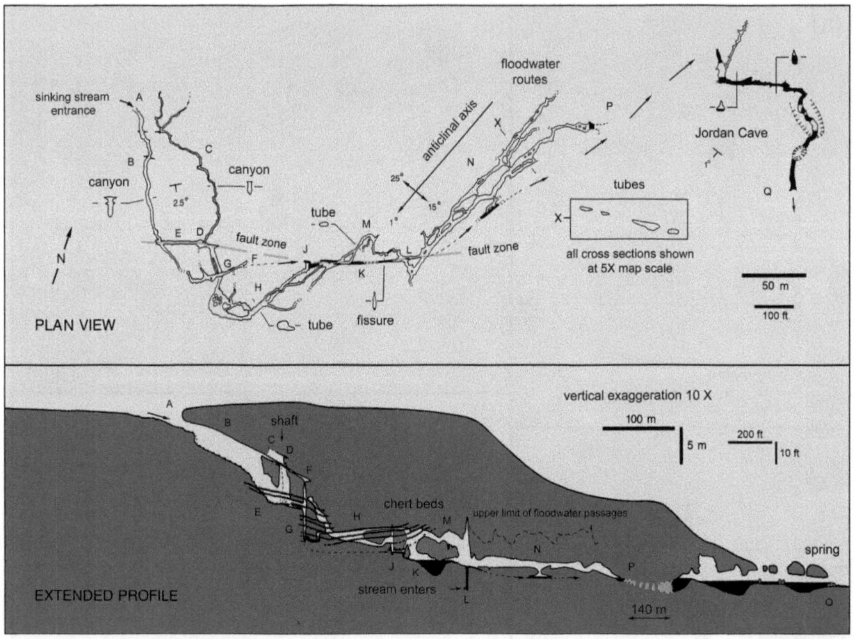

Fig. 6.4 Plan and profile view of Onesquethaw Cave, Albany County, New York, with geologic annotations of the structural zones. Non-floodwater passage follows the anticlinal axis, the dip of bedding planes, and the fault. Floodwater passages tend to ignore these structures. (Map reproduced from Engel 2009)

Fig. 6.5 Onesquethaw Cave, Albany County, New York. **a** Entrance canyon under dry conditions. **b** Entrance canyon in flood. **c** Downstream phreatic passage, showing the roof as the sloping limb of the anticline guiding this section of cave; note scalloping indicating high flow velocities. **d** Entrance to the cave in 1968 in dry conditions, where the surface stream enters the cave in flood; author Mylroie is kneeling, with three other neophyte cavers from Syracuse University. **e** Same cavers as in d, in the large room at the downstream end of the entrance canyon. (Photos **a**, **b** and **c** courtesy of A. Palmer, Peg Palmer for scale; **d** and **e** courtesy of Joan Saxon)

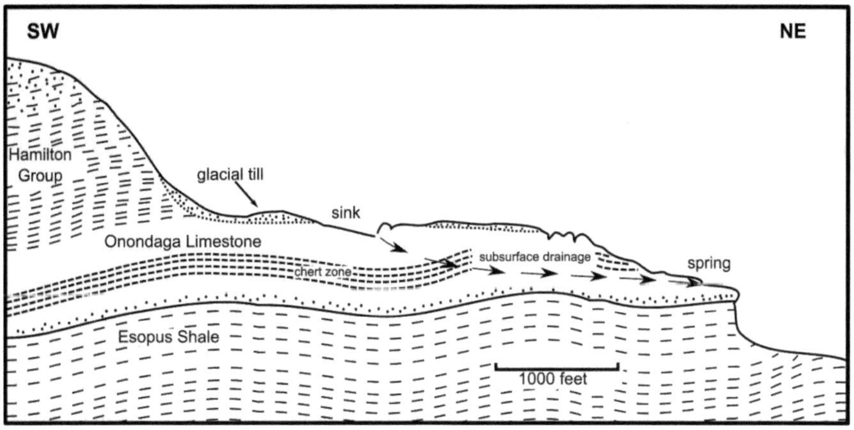

Fig. 6.6 Cross-sectional view of the geology and hydrology near Onesquethaw Cave, Albany County, New York. Water here is guided over glacial till before sinking into Onesquethaw Cave, and contributing to Onesquethaw Creek. The perching of this tributary on Wisconsinan glacial till, and no indication of control by prior, pre-Wisconsinan flow routes show that Onesquethaw Cave is potentially post-glacial in origin. (Redrawn from Palmer 1972)

were guided by a thrust fault, as well as the strike of bedding planes within an anticline. Pirated waters additionally began the formation of a downstream cave that acts as the resurgence, Jordan Cave. The deeper, fault controlled passages are what Palmer (1972) calls "ungraded passages" (equivalent of a phreatic lift tube of Ford and Ewers 1978), where phreatic passages are not level, but go up and down. Ungraded passages additionally cut through the axis of the anticline. The styles along the bedding planes are "graded" and do not go up and down, indicating formation at the water table. The anticline-breaching passage is interesting, the passage going downstream ascends the limb of the fold as a phreatic tube; once over the fold axis, the passage becomes a vadose canyon as the water spills over and runs "downhill" along the descending anticline limb to become phreatic passage again at the base of the limb, when the passage turns to follow the strike of the fold (Fig. 6.5c).

Stages three and four see a migration to vadose incision of the passages formed in stage two, as Onesquethaw Cave became efficient enough to transport the entire base flow of the sinking stream. Though this is the case, stages three and four show considerable floodwater modification, generating the anastomotic pattern seen through parts of the cave. The style of development of these two stages is very similar, with base flow incising in a vadose style, and floodwaters enlarging upper levels, with some control by structures such as jointing associated with the anticline. Floodwater morphology is seen throughout the cave, such as dead-end passages. The final stage of evolution for Onesquethaw Cave, stage four, is the migration of the insurgence point for the cave 1000 ft (305 m) upstream, leaving the previous entrance dry except during high flow (Palmer 1972) (Fig. 6.5a, b).

Of interesting note, both Clarksville Cave and Onesquethaw Cave have a variety of impacts caused by floodwaters, but of different origins. Floodwaters (and perhaps glacial meltwaters) in Clarksville Cave established maze-work that can be attributed to back-flooding from sediment plugs, as well as collapse, as commonly seen in Helderberg systems to the west, whereas Palmer (1972) attributes these features in Onesquethaw Cave to inefficiencies in downstream passage due to structure leading to the resurgence, Jordan Cave. Both of these caves show influences of structure on caves with faults guiding lower level, base flow passages. Floodwater morphology, combined with strong structural controls of non-floodwater passages show how similar cave evolution can occur in different glacial stages, and indicate that more information is needed to distinguish this evolution, such as the glacial sediment and multiple phreatic levels in Clarksville Cave. Onesquethaw Cave has been long recognized as being vulnerable to flooding during high discharge events (Fig. 6.5b). As noted earlier, the surface stream normally sinks to the west of the entrance, but overflow water goes directly into the entrance and the cave can flood to the ceiling. In December, 1990, a party of cavers was trapped in the cave by floodwaters, but successfully rescued. The cause of the flood was the human-induced breaking of a beaver dam on a lake upstream of the cave, such that floodwaters entered the cave on what was a clear-sky day (Engel 1991).

The speculation by Palmer (1972) that Onesquethaw Cave is post-glacial in origin is supported by the U/Th dating of Lauritzen and Mylroie (2000), who record the presence of only a few, small speleothems, none of which dated older than 5 ka. However, in a cave that currently undergoes severe backflooding by aggressive waters, older speleothems could have been removed or never have had an opportunity to form. Older speleothems may also be present, and not sampled by the Lauritzen and Mylroie (2000) project. Onesquethaw Cave may be post-glacial in origin, but conclusive evidence is not available.

The contrasting origin times for Clarksville and Onesquethaw caves and their styles of development are important to the interpretation of other caves in glaciated karst settings. These caves both exist in the Onondaga Limestone, have similar structural controls, surveyed lengths, and multiple levels, and are in close proximity. Despite this, these caves appear to have different origin times; Clarksville Cave is pre-glacial in origin, while Onesquethaw Cave may be post-glacial. These origin times are distinguished by different styles of development: Clarksville Cave has multiple, phreatic levels of high cross-sectional area that are established with base levels changes over multiple glaciations; Onesquethaw Cave's levels are vadose or epiphreatic during flooding, two styles of development that lead to rapid speleogenesis. Caves with multiple, phreatic levels are likely pre-glacial, or perhaps subglacial in origin (e.g. Mitchells Cave, Chap. 7), while vadose caves formed by sinking streams, even with large passage cross-sectional areas, can be established in the time since deglaciation.

6.4 Joralemon Park Karst Area, Ravena

Joralemon Park is located in southernmost Albany County, New York, on the Ravena 7.5-min quadrangle, though it only occupies a small portion of it. This park contains several caves, both pre- and post-glacial in origin. Most of the caves in this small area are contained within the park, though the boundary of the drainage basin acting as the catchment for several of the caves extends kilometers outside of the park. The area containing this catchment, the karstic portion of the park, and the portion of the park with no known karst is less than 4 km^2 (1.5 $miles^2$) (Fig. 6.7). In the literature this karst area has been called the Hannacroix Maze karst (e.g. Nardacci 1994), and the Joralemon Park karst area (e.g. Rubin et al. 1995; Cooper and Mylroie 2014). While this area is small compared to the karst area of the Helderberg Plateau, and indeed smaller than even specific cave system subsections of the Helderberg Plateau, it contains a mixture of caves of different origin times, geometries, and other characteristics.

A full geologic description of the surrounding area can be found in Goldring (1943). Fine details of hydrology, geology, and karst within Joralemon Park can be found within Nardacci (1994), Rubin et al. (1995), and Cooper and Mylroie (2014).

6.4.1 Geology and Hydrology

The karstic unit of this area is the Onondaga Limestone, which is both overlain and underlain by clastic units. Surrounding Joralemon Park, these units are highly folded (Fig. 6.1), with axes oriented predominantly north-south. These folds guide flow of the local hydrology, with ridges acting as drainage divides, valleys acting as paths for streams, and glacially carved depressions containing swamps. The clastic Esopus Formation acts as the local dissolutional base level. Quaternary deposits also exist in the form of tills that cap the ridges of the area, and a moraine that separates the larger drainage basin into two smaller basins. Large, apparently locally derived (consisting of Onondaga Limestone) erratics exist as deposits (Fig. 6.1c). Another, clearly glacial feature is a roche moutennée containing two of the caves in the area, Hannacroix Maze and Merritts Cave. The structural grain of the fold axes, oriented north-south, is parallel to the adjacent Hudson River, and to the ice flow direction (south) in this part of New York State. Therefore the structural, fluvial, and glacial grain is the same at the large scale, which makes interpreting the controls of cave development in this area a bit more difficult than in the main Helderberg Plateau to the west.

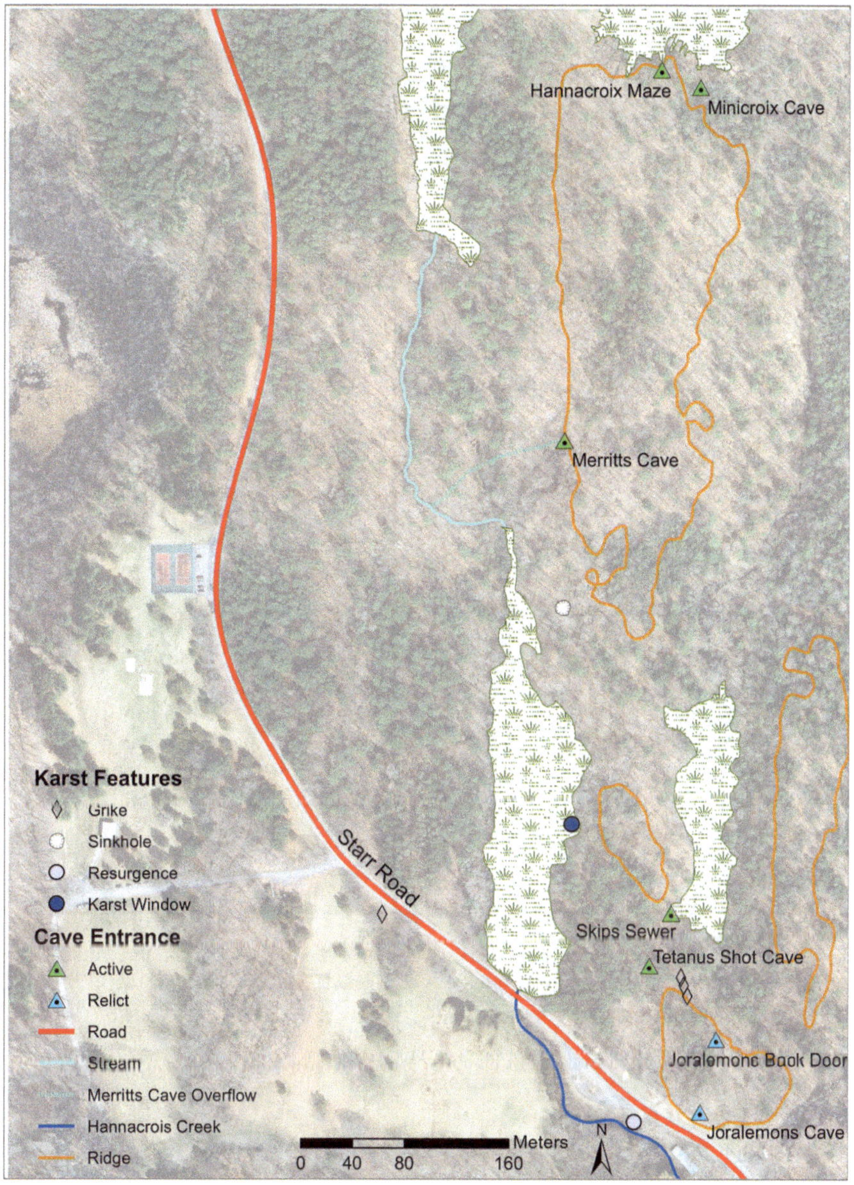

Fig. 6.7 Area map of Joralemon Park in Albany County, New York showing hydrologic, karstic, glacial, and structural features. The ridges in the park are aligned with both ice flow direction, as well as fold axes of anticlines. The hydrology of the park is complex, and is controlled by glacial features, structure, and karst. Water flows year round through the resurgence, the ultimate downstream karst feature. During low flow the surface stream adjacent to the ridge containing Hannacroix Maze is dry; Merritts Cave is also dry during low flow. During higher flows water can be found in Merritts Cave, and flow can occur out of the karst window. Even higher flows allow water to flow out over the normally dry streambed, and also out of Merritts Cave. Flow during this time out of Merritts Cave sinks back underground at the sinkhole. During the highest of flows Merritts Cave can be full to the ceiling, with water from Merritts Cave joining the surface stream

The hydrology of this area is a mixture of deranged and karst drainage. This area contains several swamps that are connected to each other by small, inefficient draining streams. An isolated, small swamp of the smaller sub-basin is apparently drained entirely through dissolutional caves, including Skips Sewer (Fig. 6.8a). Where swamps abut the outcroppings of the Onondaga Limestone flow is directed through the active caves of the area, including Hannacroix Maze (Figs. 6.8b and 6.9), and Merritts Cave (Fig. 6.8c).

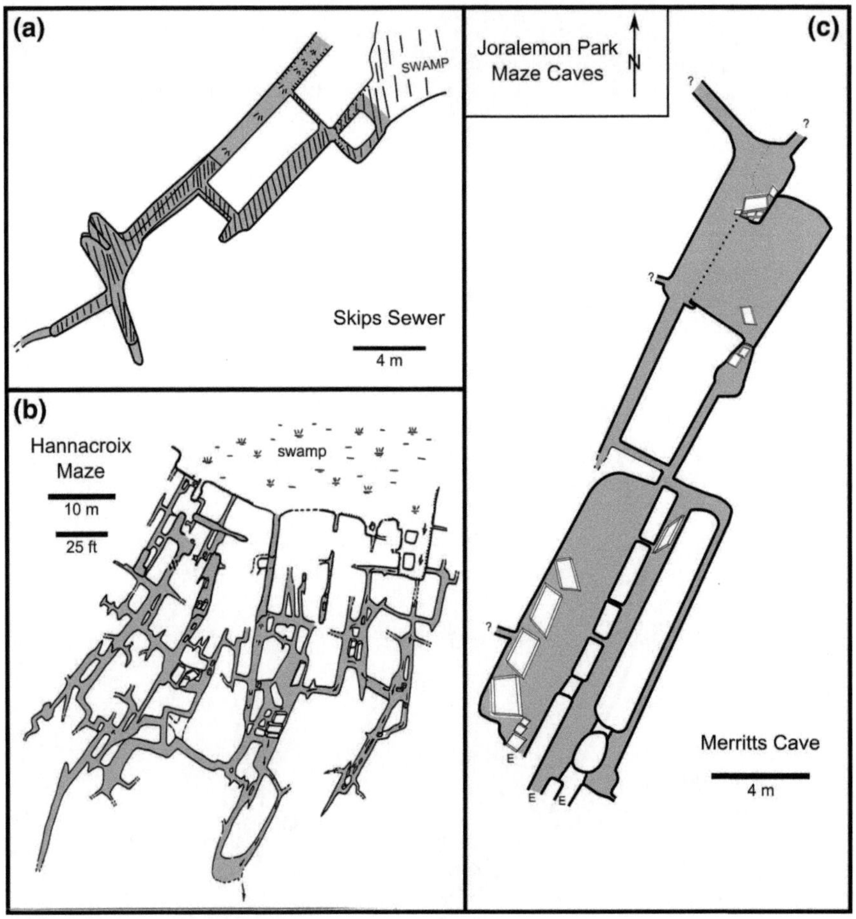

Fig. 6.8 Maps of maze caves located in Joralemon Park. **a** Skips Sewer drains a swamp from the north, and is currently occluded with organic matter. **b** Hannacroix Maze is the largest maze cave in Joralemon Park, with 2000 ft (610 m) of mapped passage, with more to be explored. (Reproduced from Palmer 2007). **c** Merritts Cave has a three-dimension aspect to it, with low-ceilinged rooms at a higher level on top of large breakdown blocks. Merritts Cave is hydrologically connected to Hannacroix Maze, and a humanly passable connection may exist between the two caves

Where no carbonate outcrop exists flow is instead routed over land on the clastic Schoharie or Esopus formations. The structural, glacial, and karst controls create a very complex drainage regime, with water that flows through cave systems rejoining the surface streams at differing points depending on discharge (Fig. 6.7). In some cases water flows out of caves, over the surface, and sinks back down into a sinkhole only to reemerge at a karst window. The ultimate resurgence for the streams and caves is Hannacrois Creek (spelled differently than the cave, for unknown reasons, Nardacci 1994). Water resurges through a karst window apparently downstream of Hannacroix and Merritts Cave, as well as the sinkhole, and through an underwater spring directly into Hannacrois Creek.

6.4.2 Caves and Karst of Joralemon Park

Joralemon Park contains a healthy mix of pre-glacial and post-glacial caves (Table 6.1). Interestingly, the majority of the caves are interpreted to be post-glacial, and have a network maze geometry, following the closely spaced jointing in the roche moutonnée, and other small ridges. Currently non-enterable conduits active at the same base level are interpreted to be post-glacial as well (Cooper and Mylroie 2014).

This area contains two, definite pre-glacial caves, Joralemons Cave (also called Fish Club Cave in the literature, e.g. Steadman et al. 1993) and Joralemons Backdoor (Fig. 6.10). These caves are relict and truncated, with entrances existing at a higher elevation

Fig. 6.9 Photograph of a room and passage in Hannacroix Maze. **a** Larger rooms in maze caves are frequently formed due to breakdown; similar large rooms in maze caves are found in nearby Merritts Cave, as well as in Barber Cave of the Helderberg Plateau (Chap. 5), Glen Park Labyrinth near Watertown, NY (Chap. 7), and X and Big Loop caves of the Adirondack Mountains (Chap. 9). **b** The passage shows a phreatic fissure type passage, a common morphology in maze caves. Chert nodules, typical of the Onondaga Formation, can be seen. (Photos courtesy of M. Chu)

Table 6.1 List of caves located within Joralemon Park, their geometry, drainage position, and time origin

Cave	Geometry	Drainage position	Time origin
Hannacroix Maze	Network Maze	Active—within deranged drainage	Post-glacial (Cooper and Mylroie 2014)
Minicroix Cave	Network Maze	Active—within deranged drainage	Post-glacial
Merritts Cave	Network Maze	Active—within deranged drainage	Post-glacial (Cooper and Mylroie 2014)
Tetanus Shot Cave	?	Active—within deranged drainage	Post-glacial
Skips Sewer	Network Maze	Active—within deranged drainage	Post-glacial
Joralemons Cave	Single passage	Relict	Pre-glacial (Steadman et al. 1993)
Joralemons Backdoor	Single passage	Relict	Pre-glacial (Steadman et al. 1993)

Referenced literature demonstrates these time origins in detail, while other time origins are inferred by position within deranged drainage

than the current, active caves (Table 6.1), and are formed within a folded and faulted ridge (Fig. 6.11). These caves contain dateable material, including paleontological remains of an early Holocene black bear; pollen stratigraphy of sediments within the caves also indicates that the caves were enlarged pre-Holocene (Steadman et al. 1993).

These two pre-glacial caves have the cross-sectional geometries of phreatic, master trunk passages. Their orientations suggest that these two caves are connected, but this connection, and the true dissolutional floors of these caves, are occluded by glacial sediment. Their position and cross-sectional size indicate that they once drained a substantial, perhaps branchwork pre-glacial cave system. Scallops indicate southward flow, similar to that of the current drainage regime. These scallops indicate slower velocities than those of the floodwater-formed maze caves (but with a larger passage cross section, could indicate a higher overall discharge).

From these two pre-glacial caves a picture can be painted of an earlier, larger cave system, with waters coming from the north off of folded ridges, flowing through a perhaps dendritic, branchwork cave, and continuing towards the south towards the pre-glacial base level. As noted for Onesquethaw Creek, some streams may survive glaciation (Dineen 1987), and Hannacrois Creek may be one such example, thus being the pre-glacial base level control. One can envision vadose waters flowing east and west down dip of anticlinal limbs, and into phreatic, strike oriented passage. As these caves are relict, and their surrounding area glacially modified (e.g. roche moutonnée), it is difficult to paint an exact story of a

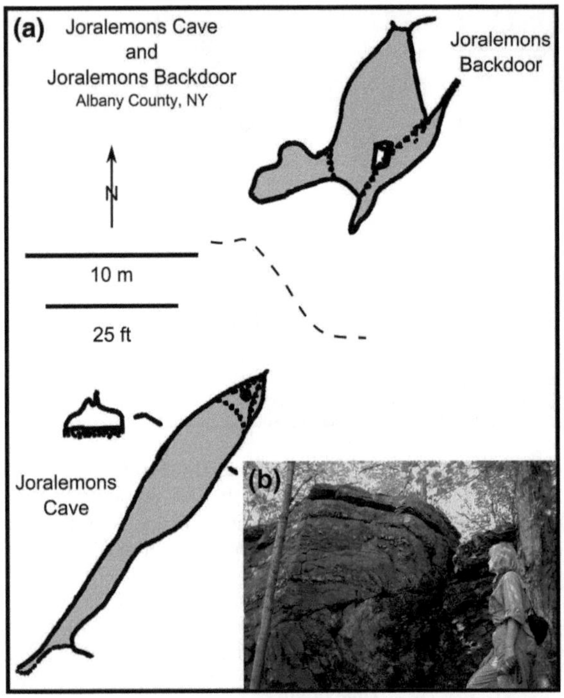

Fig. 6.10 Map of Joralemons Cave and Joralemons Backdoor (**a**). It is inferred that these two caves are part of the same master trunk passage, as they are aligned, though their connection is occluded by glacial sediment. The orientation of these caves is guided by faulting and folding (**b**). In map (**a**) distance is not preserved between Joralemons Cave and Joralemons Backdoor, though orientation and alignment is; for a map showing the accurate distance see Rubin et al. (1995). (Original cartography by T. Engel; modified from Rubin et al. 1995). Photo (**b**) is taken at the entrance of Joralemons Cave

history where insurgences and resurgences were plugged, thusly reorganizing flow routes, though ultimately preserving pre-glacial flow routes, such as presented by larger Helderberg caves to the west and their survival through multiple glaciations. The small area at Joralemon Park, the presence of active caves at the current base level, the glacial erosional landforms, and the organization of surficial glacial deposits do allow a nice story of the transition of pre-glacial cave systems to different modern cave systems.

The active caves of Joralemon Park do not display any change in base level through multiple levels, indicating that they are possibly post-glacial in origin. Additionally these caves can form in the time since deglaciation (Cooper and Mylroie 2014) based on the high wall-retreat rates generated when forming maze caves under floodwater conditions. All of the caves are additionally shallow, with ceilings existing within three or less meters from the current surface. Adjacent plucked blocks of Onondaga Limestone prominences are of greater dimension than the caves are deep. If these caves existed pre-glacially, why did plucking stop when there is clear weakness (open cave passages) within a few meters? All these pieces of evidence lend to post-glacial origins of these caves, perhaps with some initial sub-glacial initiation.

The surficial glacial deposits aid in understanding the evolution of the maze caves in Joralemon Park. As the glaciers retreated moraines were left in Joralemon Park, and in its vicinity. One such moraine divides the upper swamps from the swamp that contains Skips Sewer. The glacially carved, and structurally and glacially aligned ridges and valleys are coated in glacial till. In Hannacroix Maze this till can be seen in the Sleeping Alligator entrance, where floodwaters have emplaced cobbles, and where dissolution has proceeded upwards to the till cap overlying the roche moutonnée. The arrangement of depositional material has allowed the clear separation of surficial catchment areas, particularly in low flow regimes, guiding the evolution of active caves at current base level. Of note is the moraine separating the two catchment areas, as Rubin et al. (1995), following Dineen (1986) claim that the elevation of this moraine correlates with that of a lake level of Lake Albany, 335 ft (100 m) amsl. This moraine also provides a drainage divide, leaving Skips Sewer and any downstream passage to evolve on its own, not coupled to the other post-glacial caves of the area.

If the interpretation of Rubin et al. (1995) is correct, this area may have been inhabited by Lake Albany. If this were the case (and even the case with a smaller lake), a large hydraulic gradient would exist along the Onondaga Limestone outcrop. Combined with mechanically enlarged joints, high hydraulic gradients would generate rapid, post-glacial breakthrough, and inundate an early developing Hannacroix Maze. Rubin et al. (1995) argue, based on the idea that it takes 10,000 years to form a humanly traversable cave, the time since deglaciation is not enough to enlarge this cave to its current dimension; thus Hannacroix Maze initially formed sub-glacially, followed by post-glacial enlargement. Cooper and Mylroie (2014), however, describe this cave as entirely post-glacial, and demonstrated that the inundation of a lake (argued by Rubin et al. 1995), followed by high wall-retreat rates due to floodwaters could form the cave in the time since deglaciation without invoking sub-glacial

Fig. 6.11 Photographs in and near Joralemons Cave and Joralemons Back Door Cave, Albany County, New York. **a** Joralemons Cave Entrance, near the top of a low ridge, looking north. **b** Joralemons Back Door Cave entrance, looking south (the two caves do not actually connect; sediment obstructs the linking passage). **c** Inside Joralemons Cave, looking north in the paleo-upstream direction. **d** Inside Joralemons Cave, looking south and downstream to the entrance. The large passage size, and position well above base level in an isolated hill, plus Late Pleistocene paleontological evidence indicate a pre-glacial age for the cave

speleogenesis. Mechanical enlargement due to breakdown and subsequent transport or dissolution of breakdown blocks (Fig. 6.9), or abrasion by material during flooding also aids in rapid cave enlargement.

The roche moutonnée gives an additional hint on post-glacial karst development. As noted in Chap. 4, talus caves can form in the split side of a roche moutonnée. The talus entrance (Fig. 6.12), and some larger, tilted blocks that produce a three dimensional maze-form to Merritts Cave indicate its life may have begun partially as a talus cave, albeit one without large passage. While this may or may not be the case, there is clear dissolutional enlargement of Merritts Cave, including scallops. With no extra catchment area, and

Fig. 6.12 Entrances to Merritts Cave (**a**), and Skips Sewer (**b**). The entrance to Merritts Cave exists within a talus pile on the western side of the roche moutonée in Joralemon Park. The entrances to Skips Sewer are located within dissolutionally enlarged joints (grikes). Hannacroix Maze has similar entrances (though larger in dimension) to Skips Sewer

no evidence of substantial vadose, downward flow into Merritts Cave, this cave is interpreted to be a downstream continuation of Hannacroix Maze. If this connection is the case, then there is still much passage to be discovered between them. It also presents the question: how extensive is the maze-work? If the maze passages continue between these two caves, there is no issue. If the connection is through a single conduit though, how did the second, downstream maze, Merritts Cave, form? Two hypotheses for this are: the initial origin of this cave is indeed from a split roche moutonnée, or there is some inefficiency raising hydraulic head, thus causing complete inundation of joints in Merritts Cave. Either case produces an interesting story for speleogenesis, both with some tie to glaciation.

An important issue in the discussion about subglacial versus post-glacial origin for some of the Joralemon Park caves is that even within Holocene time, significant changes have occurred in the drainage basin. What were initially small lakes have become swamps as fluvial infilling has occurred. Barriers that separate lakes and swamps, especially if made of glacial till or moraine material, can be downcut. These activities change the amount of water storage, and the amplitude and duration of high discharge events within such a drainage basin. One must be careful about extrapolating current discharge characteristics of these drainage basins backwards in time to the start of the Holocene. Cooper and Mylroie (2014) address these issues when discussing supposed post-glacial maze caves both here and elsewhere in New York.

6.5 Pompeys Cave

Pompeys Cave (Fig. 6.13) is the largest dissolutional cave in Ulster County, New York, with a surveyed length of 1148 m (3788 ft). Unlike the caves mentioned previously this cave formed in the Rondout Formation (Late Silurian in this location), a minor karst former elsewhere in the Helderbergs. Particularly this cave formed in the Glasco Limestone, a member of the Rondout Formation not present to the west. The Glasco is underlain by the Rosendale Member and overlain by the Whiteport Member, both dolomites. The presence of this limestone member is probably the reason this cave is the largest known from the Rondout Formation in New York. This cave accepts the entirety of Kripplebush Creek under low flow (Figs. 6.14 and 6.15a), and only during high flow does the creek flood the cave (Figs. 6.15b, d and 6.16a) and flow over the Whiteport Member (Fig. 6.16b). Kripplebush Creek fits within the current, Holocene deranged drainage with upstream waters flowing through several ponds and swamps located in glacial depressions.

This cave has been described and interpreted in relation to structural controls on the cave development by Bowles and Bartholomew (2013), who also performed a ground penetrating radar (GPR) study on the downstream, sumped passage (Bowles et al. 2014). While this cave has been interpreted in relation to its structural controls on cave development and evolution, there has not been work to interpret the history of the cave in relation to glaciation. Using the perspective provided by other studies related to glaciation in the karst of the HVFTB, an interpretation is made here to

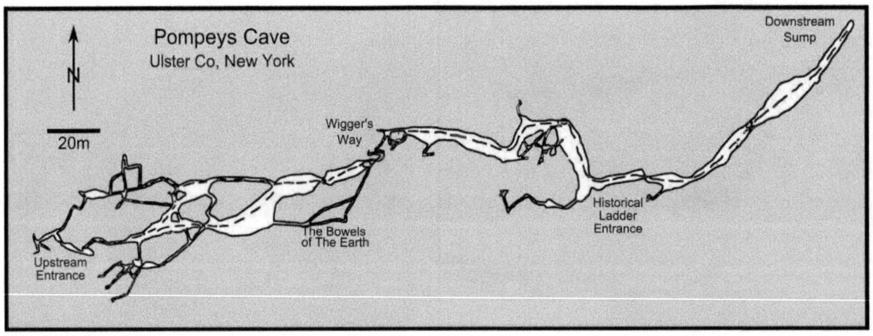

Fig. 6.13 Map of Pompeys Cave in Ulster County, New York. This cave is mainly a trunk stream passage, with some floodwater maze-work passages in the upstream portion of the cave. This maze-work formed due to backflooding due to constriction of Wigger's Way by breakdown and structure. (Modified from DeThomas et al. 1988)

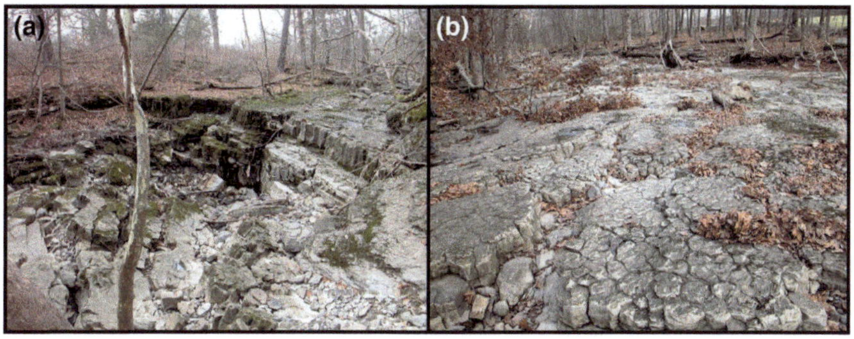

Fig. 6.14 Kripplebush Creek at Pompeys Cave, Ulster County, New York. **a** Dry Kripplebush Creek, with a small waterfall formed by collapse into the cave, at the center and left of the image. **b** Dry bed of Kripplebush Creek, displaying fossil mudcracks in the Rondout Formation. (Photos courtesy M. Chu)

Fig. 6.15 Photographs from inside Pompeys Cave, Ulster County, New York. **a** Main stream passage downstream of the Ladder Entrance (Fig. 6.13). **b** Main stream passage in flood, note small waterfall on the left. **c** Collapse and flood debris in the main stream passage. **d** Collapse material with embedded flood debris. (Photos courtesy M. Chu)

its time origin, and evolution controlled by structure, glaciation, and floodwaters.

An interesting note on Pompeys Cave is the origin of the name, and a history going back into at least the 1800s. Knowledge of the cave has existed since the 1800s and may have appeared in a local newspaper in 1832, as the description matches Pompeys Cave (DeArmond 1969). The name Pompeys Cave originates from a slave named Pompey who would go down into the cave during the summer to keep cool.

Several other caves are located near Pompeys Cave including Pompeys Annex, Pit Cave, and Fossiliferous Cave. These caves are formed in the Manlius Formation located stratigraphically above the Rondout Formation.

6.5.1 Geology and Hydrology of Pompeys Cave

As stated above Pompeys Cave is formed in the Glasco Limestone member of the Rondout Formation, bounded on top by the Whiteport Dolostone, and on the bottom by the Rosendale Dolostone. The dolostone members are relatively less soluble, with the

Fig. 6.16 Photographs of Pompeys Cave during flooding conditions. During large discharge events Pompeys Cave can fill entirely; when Pompeys Cave is full the normally dry creek bed can become active. **a** Photograph taken from ladder in the Ladder Entrance to Pompeys Cave. Here the cave is not quite in conduit-full conditions. **b** Water flowing over the normally dry creek bed. The ladder entrance (under the diagonal slab) here is completely inundated. Note the large tree trunk being transported by the creek; contributions of mechanical weathering in high flow conditions can contribute to high rates of speleogenesis. (Photos courtesy A. Bartholomew)

Whiteport Member guiding flow overland until it sinks upon contact with the underlying Glasco Member. The Rosendale Member acts as the dissolutional base level. The Glasco Limestone within Pompeys Cave contains a biostrome in some locations. These units are folded with most of the cave formed within the core of a plunging anticline, and in places under the hinge of a monocline guided over a biostrome (Bowles and Bartholomew 2013). Joints associated with the formation of the anticline also guide development of smaller passages in the upstream portion of the cave.

Flow into Pompeys Cave is directed overland in Kripplebush Creek by the Whiteport Dolostone. Only when flow reaches the contact with the Glasco does it sink into several swallets within the creek bed. Several larger swallets act as entrances to the cave (Figs. 6.14a and 6.16b), including the historic Ladder Entrance of the downstream section of the cave. After large flood events these swallets can become occluded with organic debris, including entire tree trunks, only to be cleared out again by another flood event. Pompeys Cave also has multiple resurgences, including smaller, diffuse ones. These resurgences are connected to the cave via sumped passage, with all upstream passage being either of vadose, or floodwater origin. The two major resurgences are the master resurgence, and the overflow resurgence (Fig. 6.17). During large flow events, such as during snowmelt, the overflow resurgence yields great amounts of water. Large flood events also cause water to flow over the cave entirely as it fills with water. Flow within the creek bed can be spectacularly raised, with changes in bed topography of over three meters being completely inundated so that only a small waterfall occurs where under lower discharges, the water fall is 3 m high.

The non-sumped portion of Pompeys Cave is constrained to one, actively forming vadose level. In both the upstream and downstream sections of the cave the larger passages contain year-round water. The upstream portion of the cave contains two active trunk passages, and smaller passages that connect these, some of which do not contain year-round water. The upstream portion of the lower cave ends in constricted passage, Wigger's Way (Fig. 6.13), which connects to the upper portion of the cave. Wigger's Way contains breakdown blocks and is 0.35 m (1.1 ft) wide. Downstream, the cave is mainly limited to a single trunk stream passage, with only a few small side passages. Bowles and Bartholomew (2013) note that for every narrow passage in the downstream section of the cave there are five in the upstream portion in a maze-work pattern. As the connection between the upstream and downstream portions of the cave is constricted by geology, breakdown, or a combination of both, the upstream smaller side passages represent a backflood maze superimposed on the cave.

At the ultimate downstream portion of Pompeys Cave is a sump. This sump continues downstream and resurges back to the surface after ~100 m (~330 ft) at the contact between the Glasco Member and the Rosendale Member in several locations (Bowles et al. 2014). Of interest are the two major resurgences: the

Fig. 6.17 Photographs of the master resurgence (**a** and **c**), and overflow resurgence (**b** and **d**) of Pompeys Cave. **a** Master resurgence during high flow conditions. **b** During high flow conditions water flows out of the overflow resurgence. **c** Master resurgence during low flow conditions. Water flows out of this resurgence year-round as base flow. **d** The overflow resurgence during low flow is completely dry. (Photos **a** and **b** courtesy A. Bartholomew; photos **c** and **d** courtesy M. Chu)

master resurgence and the overflow resurgence (Fig. 6.17). As one only yields water during high flow, some restriction to flow must occur between the main route and the master resurgence. Work with GPR reveals shallow, several meters wide passage with dissolutional features. This passage apparently contains some amount of breakdown material (Bowles et al. 2014).

Like the previous caves of this chapter Pompeys Cave does contain glacial sediment. The presence of this sediment allows a question to be asked: was this sediment deposited by glaciers, and then partially removed by high flows, or were these sediments deposited post-glacially with high flows transporting them in allogenically?

6.5.2 Evolution of Pompeys Cave

Unlike the previous caves and cave areas of this chapter, Pompeys Cave seems simple to interpret upon first glance. This cave is very similar to Onesquethaw Cave; the cave accepts an entire sinking stream, and only during high flow events does flow over the surface occur due to inefficiency in the subsurface route caused by a constriction. Also like Onesquethaw and Clarksville caves, Pompeys Cave has the superposition of a floodwater maze onto master trunk passages. The inefficiency at Wigger's Way part way through the cave (Fig. 6.13) causes backflooding into joints and bedding planes upstream, thus forming maze-work passages that ignore the guiding influence of the anticline. Joint control on several of the maze passages such as the Bowels of the Earth section may have been aided in their initial formation by mechanical expansion during post-glacial rebound.

Unlike Onesquethaw and Clarksville caves, Pompeys Cave exists at entirely one level. The limited vertical extent of the Glasco Limestone (~5 m) is partially the culprit of this morphologic difference. The other guiding factor for a single level may be that the cave is entirely post-glacial in origin. Kripplebush Creek, and thus Pompeys Cave, fit within the Holocene deranged drainage, with no indication of passage following any pre-glacial flow routes. The enterable, vadose section of the cave, as well as the sumped passage exists as one, active level (though it may be the case that this is due to the perching on an insoluble unit). Glacial sediments within the cave can easily have been transported allogenically as massive flooding occurs regularly in this cave. This cave therefore appears to satisfy the criteria of a post-glacial, Holocene cave laid out in Mylroie and Carew (1987), and Cooper and Mylroie (2014). A post-glacial origin for this cave then gives a starting point to examine its evolution, though other interpretations can be made such as this cave being pre-glacial, guiding the current Holocene drainage.

The Manlius Formation in the area around Kripplebush Creek is present in limited outcroppings, with only small caves within it. The Rondout Formation in this area is more massively bedded, including both the dolomite and limestone units. The presence of the Glasco Limestone member of the Rondout Formation allows for cave formation in a unit that is typically not karstified. The thin-bedded Manlius Formation may have been preferentially removed by repeated glacial scour through the Quaternary, removing any large, pre-glacial systems within it at this location, if such existed here.

Initially during and after deglaciation it is likely that flow took place over the now mostly dry Kripplebush Creek bed. As dissolution proceeded at the contact of the Whiteport and Glasco members flow was pirated into the subsurface, eventually to accept the entire low-flow stream. During this time the initial phreatic development occurred, likely establishing what is the currently sumped passage, and the beginnings of the main trunk passage. As more flow was captured to the subsurface, the stream was diverted underground and vadose entrenchment occurred in the upstream and downstream portions of the cave, with continued phreatic development of the sumped passage. At some point in this evolution the inefficiency at Wigger's Way developed, probably as a result of greater passage size inducing a major collapse event, allowing the epiphreatic floodwater maze to be superimposed on the mostly vadose stream passage. A hypothesis offered by Bowles and Bartholomew (2013) is that this restriction is from a fold anomaly at this location. If this anomaly is the case, the combination of it and breakdown between the upstream and downstream portions of the cave allowed the formation of this maze. If this anomaly is not the case, breakdown alone may be the cause of the inefficiency. A similar inefficiency must be present downstream as well, causing the overflow route to develop during high discharge events.

The single level of Pompeys Cave indicates that the developmental history is simpler than that of the similar Onesquethaw Cave. Pompeys Cave continues to develop in the same fashion after its initial development, with multiple active sinking stream inputs guided through a cave stream. No apparent fossil passage exists to tell a story involving diversions away from previous inputs, like that of Onesquethaw Cave. The single most important aspect of the interpretation of Pompeys Cave as post-glacial is that the main flow route is quite large, commonly several meters high and 3–6 m wide. This is a large passage cross section to form in $\sim$10,000 years, equivalent to large passages found in pre-glacial caves such as Clarksville Cave. Kripplebush Creek is a much larger stream than the one that sinks at Onesquethaw Cave, and the large size of Pompeys Cave may be only a result of the increased scale of discharge. The large amount of coarse sediment load the cave transports may also allow a degree of mechanical erosion not found in smaller post-glacial caves.

6.6 Other Caves of the HVFTB

As fewer studies have been performed within the HVFTB karst compared to the Helderberg Plateau, the caves presented here are not an exhaustive list of caves of this region. The systems within the HVFTB have limited catchment, and limited outcropping as a result of the interplay between structure and topography. Glaciation has deranged the surface drainage, scoured limestone outcrops, and buried others under till, further complicating the understanding of speleogenesis in the HVFTB. Despite this, other caves in this area have formed, such as Mystery Cave (also known as Surprise Cave; Fig. 6.18), the longest cave in the HVFTB with a length of 3 km (1.9 miles), and a vertical extent of 52 m (170 ft) (Engel 2009). Caves other than those detailed above have received little attention as to their speleogenesis.

Southern Albany County contains numerous caves, some of which are widely known, such as South Bethlehem Cave in the Coeymans and Manlius Limestones, Wiltseys in the Coeymans and Kalkberg Limestones, and the Camp Allen Caves in the Becraft Limestone. Greene County is known for Railroad and Lizard Cave in the Manlius Limestone, and Jack Packers Cauterskill Cave in the Onondaga Limestone. These caves are generally less than a kilometer in length, commonly only a hundred meters or so total. They reflect the small outcrops in which they occur, as well as the variety of structural controls present, which leads to cave and karst involvement of all the

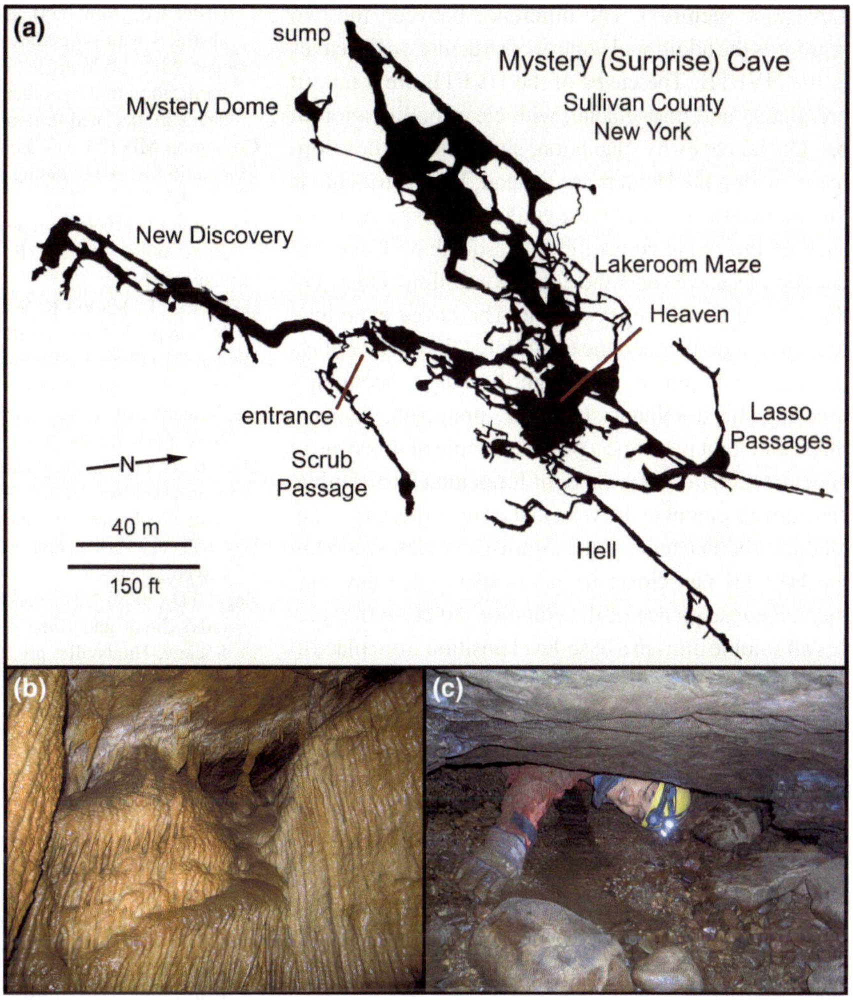

Fig. 6.18 Mystery (Surprise) Cave, Sullivan County, New York. **a** Map of the cave **b** Massive flowstone, perhaps evidence of a pre-glacial origin. **c** Glacial cobbles and gravels, typical of caves in the northeast, even though this cave is near the southern glacial limit

Helderberg and Onondaga rocks in this area (Rondout, Manlius, Coeymans, Kalkberg, Becraft and Onondaga). For these units, stratigraphic position is less of a factor than in the flat-lying limestones of the Helderberg Plateau to the west, and all of these rock units can end up hosting base-level flow systems.

Hollyhock Hollow Cave is another cave located in the HVFTB, located to the southeast of Onesquethaw Cave, and is one cave that has received attention in the literature on glaciated karst. Lauritzen and Mylroie (2000) dated speleothems collected from this cave, showing growth at ~70 ka BP from three samples and ~40 ka BP from four samples. The samples were collected from sinkhole debris material created by digging to find a cave, they were therefore not in situ. These dates bracket the Mid-Wisconsonin interstadial (MIS 3). As Hollyhock Hollow was the southernmost cave location used in this study, they related the 40 and 70 ka dates to ice advance and retreat, which created a larger time window for speleothem growth than in other caves of their study farther to the north, which saw ice retreat later, and ice re-advance sooner, than at Hollyhock Hollow (Chap 3, Fig. 3.3). While this cave is featured in the glaciated karst literature, details of its speleogenesis relating to glaciation are not presented. The speleothem dates from Hollyhock Hollow do demonstrate that this cave is pre-Holocene in origin.

6.7 Summary

The caves that develop within the Hudson Valley Fold-Thrust Belt form in the same geologic units as those in the Helderberg Plateau (with the exception of the Glasco

Limestone member). The difference between the two regions is the additional control of structure on the caves in the HVFTB. The caves of the HVFTB are a mix of pre-glacial and post-glacial, with clear modification of pre-glacial caves by glaciation, similar to the developments within the Helderberg Plateau. When structure is thrown into the mix caves are guided by these structures, such as the faults controlling Onesquethaw Cave and Clarksville Cave, and the folds controlling Onesquethaw Cave and Pompeys Cave. The caves examined here also demonstrated the control that floodwater plays in cave development, with high discharges seemingly ignoring the dominant structures, preferring bedding plane and joint partings; a prime example of this control is seen within the maze caves of Joralemon Park. Unlike cave development to the west in the Helderberg Plateau, all the carbonate units of the Siluro-Devonian section in the HVFTB participate in independent cave development, a consequence of the complex structure that places all soluble units in a base-level position from place to place.

A key comparison of the HVFTB caves with those of the Helderberg Plateau is the existence of overflow routes. In the case of the Helderberg Plateau caves these commonly form as a response to glacial sedimentation, and are thus indicators of glacial modification. In the case of the caves in the Hudson Valley Fold-Thrust Belt overflow routes are seemingly controlled by constrictions that are due to structure and breakdown, rather than some glacial modification. Noting the difference between the two is important in understanding speleogenetic evolution in glaciated environments, and thus careful noting of structure or the presence of sediments is needed to understand this evolution. Using the tools developed from both the Helderberg Plateau, and the HVFTB, allows an understanding to be built on interpreting speleogenesis and glaciation in other areas of New York, New England, and indeed other glaciated karst regions in the US and around the world.

References

Bowles EK, Bartholomew AJ (2013) Controls on passage formation in Pompey's Cave, Ulster Co., New York State. In: Geological Society of America, abstracts with programs, vol 45, no 7, p 354

Bowles EK, Rayburn JA, Bartholomew AJ, Garin C (2014) Interpreting controls on processes of cave formation in enterable and non-enterable passages in Pompey's Cave, Ulster Co., New York. In: Geological Society of America, abstracts with programs, vol 46, no 2, p 54

Cooper MP, Mylroie JE (2014) Post-glacial speleogenesis: verification of a hypothetical model, and the origins of maze caves in glaciated terrains. Cave Karst Sci 41(2):84–95

Covington MD (2014) Calcite dissolution under turbulent flow conditions: a remaining conundrum. Acta Carsologica 43 (1):195–202

DeArmond D (1969) Pompey's 1832 and Mount Eve Chamber, from White (1832) History of New York. Northeastern Caver 1(5):63

DeThomas L, Marvin R, Smyth S, Henry W (1988) A summer in Pompey's Cave. Northeastern Caver 19(1):3–6

Dineen RJ (1986) Deglaciation of the Hudson Valley between Hyde Park and Albany, New York. In: Cadwell DH (ed) The Wisconsinan stage of the first geological district, Eastern New York, pp 89–108

Dineen RJ (1987) Preglacial and postglacial drainage of the central Hudson Valley. In: Field trip guidebook, New York State Geological Association, New York, pp B-1–B-30

Engel TD (1991) The Onesquethaw Cave incident. Northeastern Caver 22(1):6–9

Engel TD (2009) New York State. In: Palmer AN, Palmer MV (eds) Caves and Karst of the USA. National Speleological Society, Huntsville, pp 27–32

Faulkner T (2009) Speleogenesis in New England marble caves. In: Proceedings of the Fifteenth International Speleological Congress, Kerrville, vol 2, pp 855–862

Ford DC, Ewers RO (1978) The development of limestone cave systems in the dimensions of length and depth. Int J Spel 10:213–244

Goldring W (1943) Geology of the Coxsackie Quadrangle, New York State Museum Bulletin, New York 332:374

Isachsen YW, Landing E, Lauber JM et al (2000) Geology of New York: a simplified account, 2nd edn. New York State Museum, Albany

Kastning EH (1975) Cavern development in the Helderberg Plateau, East-Central New York. Bulletin I of the New York Cave Survey, New York

Lauritzen S-E, Mylroie JE (2000) Results of a speleothem U/Th dating reconnaissance from the Helderberg Plateau. New York. J Cave Karst Stud 62(1):20–26

Marshak S (1986) Structure and tectonics of the Hudson Valley fold-thrust belt, New York. Geol Soc Am Bull Memoir 158:161–176

Mylroie JE (1977) Speleogenesis and karst geomorphology of the Helderberg Plateau, Schoharie County. Bulletin II of the New York Cave Survey, New York

Mylroie JE, Carew J (1987) Field evidence for the minimum time for speleogenesis. Nat Spel Soc Bull 49(2):67–72

Nardacci M (ed) (1991) Guide to the caves and karst of the northeast. National Speleological Society, Huntsville

Nardacci M (1994) The hydrology of the Hannacroix Maze Karst; Albany County, New York. Northeastern Caver 25 (3):75–83

Palmer AN (1972) Dynamics of a sinking stream system: Onesquethaw Cave, New York. Natl Speleol Soc Bull 34:89–110

Palmer AN (2007) Cave geology. Cave Books, Dayton

Quick PG (2010) Vermont caves: a geological and historical guide. National Speleological Society, Huntsville

Rubin PA (1991a) Modification of preglacial caves by glacial meltwater invasion in East-Central New York. In: Kastning EH, Kastning KM (eds) Appalachian karst: proceedings of the Appalachian Karst Symposium, Radford Virginia

Rubin PA (1991b) Flow characteristics and scallop-forming hydraulics within the Mill Pond karst basin, East-Central New York. In: Kastning EH, Kastning KM (eds) Appalachian karst: proceedings of the Appalachian Karst Symposium, Radford Virginia

Rubin PA, Engel T, Nardacci M (1995) Geomorphology, paleoclimatology and land use considerations of a glaciated karst terrain; Albany County, New York. In: Garver JI, Smith JA (eds) Field trips for the 67th annual meeting of the New York State Geological Association, pp 81–107

Ruedemann R (1930) Geology of the capital district (Albany, Cohoes, Troy and Schenectady quadrangles). New York State Museum Bulletin 285:218

Steadman DW, Craig LJ, Engel T (1993) Late Pleistocene and Holocene vertebrates from Joralemon's (Fish Club) cave, Albany County. Bulletin of the New York State Archaeological Association, New York

7 Caves and Karst of the Northern New York Lowlands

Abstract

Flat-lying Ordovician limestones form a band around the Adirondack Mountains, separating the Grenville marble caves of those mountains from the caves developed in Cambro-Ordovician marbles to the east and southeast, and from caves in Siluro-Devonian carbonates to the south (deformed) and southwest (undeformed). This cave region is understudied and little scientific work has been done, although a limestone pavement project and a maze cave examination have been published. Caves of small size, in agreement with post-glacial deranged drainage (Schroeders Pants Cave, Houghs Cave), have been identified, as well as caves of possible pre-glacial or englacial origin (Mitchells Cave). The major cave features of this region are the large and extensive maze caves along the Black and Perch rivers, multi-kilometer systems of post-glacial origin, the largest of which, the Glen Park Labyrinth, is now relict and abandoned. These caves, in tune with the post-glacial deranged drainage of major rivers, indicate that high, sustained discharges can create large maze caves within a portion of the post-glacial time window. Houghs Cave has historical importance as a stop on the Underground Railway for escaped slaves in the pre Civil War era.

7.1 Introduction

The northern New York lowlands are an area of limited relief. These lowlands follow the Great Lakes, the St. Lawrence River, the Mohawk River, and Lake Champlain basins (Fig. 7.1), and extend into northwestern Vermont. To the east and southeast these lowlands are bordered by the metamorphics of the Adirondacks Mountains and of New England. To the southwest the northern lowlands are bordered by the Allegheny Plateau, and in some areas abut the Helderberg Escarpment (Chap. 5). To the west and north, Lake Ontario and the St. Lawrence River form the boundaries. The geology of this area consists of relatively non-deformed Cambro-Ordovician strata, underlain by Precambrian Grenville metamorphics. The geology and hydrology of the region has been well described, including early descriptions and interpretations published in the *New York State Museum Bulletin* (e.g. Cushing et al. 1910). The early work in this area by Cushing et al. (1910) includes descriptions of karst including several limestone pavements in Jefferson County, as well as underground drainage routes. The modern standard reference is Isachsen et al. (2000).

The structure of the Cambro-Ordovician units in this area is similar to that of the Helderberg Plateau, with strata of relatively low dip and few deformational

M.P. Cooper and J.E. Mylroie, *Glaciation and Speleogenesis*,
Cave and Karst Systems of the World, DOI 10.1007/978-3-319-16534-9_7,

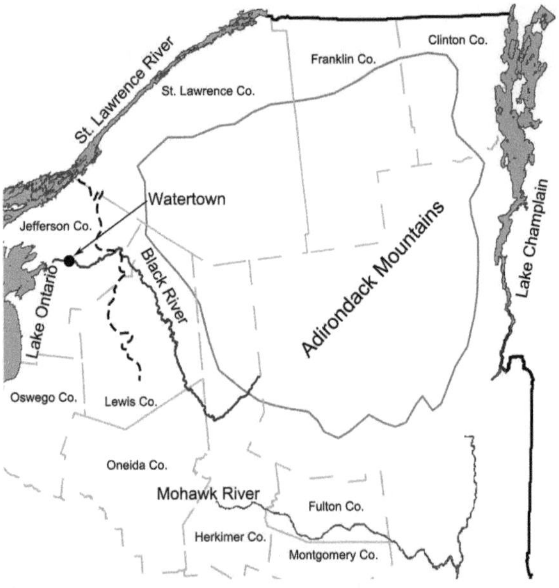

Fig. 7.1 Map of the Northern New York Lowlands showing hydrology and counties within the region. This area is bounded by the St. Lawrence River to the north, Appalachian Plateaus to the south, and New England to the east; the lowlands also wrap around the Adirondack Mountains to the north. The Black River plays an important part in speleogenesis in the region, as large maze caves are controlled by floodwaters from this river. A similar situation occurs on the Perch River, located to the north of the Black River but not pictured here. The *dashed line* indicates a pre-glacial drainage divide between the St. Lawrence River and Lake Ontario, defined by the Frontenac Axis. Breaching of this drainage divide by glaciation allowed the Black River to take an abrupt turn to the west near Great Bend, New York; to the west of this divide the Black River exists in a post-glacial channel

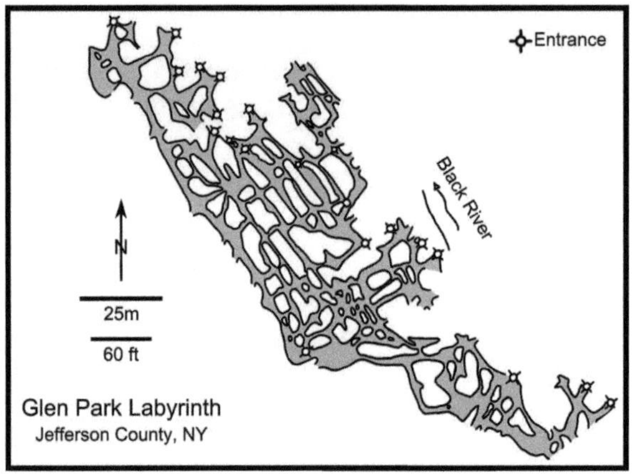

Fig. 7.2 Map of upper level of Glen Park Labyrinth. This upper level over 2 km of the 4+ km total surveyed length of the cave, and is located 18 m above the current position of the Black River in this area. Passage orientation is guided by prominent jointing in the Paleozoic strata of the northern New York lowlands, with one set trending NW-SE, and an orthogonal set trending NE-SW. Lower levels of this cave have a similar maze pattern. (Modified from Carroll 1972b)

structures, besides prominent jointing (though the exact orientations of the structures vary between the two regions). Though this is the case, the morphologies of large, extensive caves, and indeed possibly their origin times differ between the two regions.

Unlike the caves and karst of the Helderberg Plateau (Chap. 5) and the Hudson Valley Fold-Thrust Belt (HVFTB, Chap. 6) those of the lowlands have made less appearance, by number of publications, in scientific literature. The maze caves of the lowlands do however appear in one of the most influential and cited papers on maze caves (Palmer 1975), the Glen Park Labyrinth (Fig. 7.2). This cave again made an appearance in the study on the time origins of maze caves in glaciated areas by Cooper and Mylroie (2014), attributing post-glacial origins to this cave, and other maze caves in glaciated environments. A study of limestone pavements near Watertown by Feeney (1996) is one of the few such investigations from the entire northeast (Fig. 7.3).

While little attention has been given towards this region scientifically, efforts by cavers have mapped significant amounts of cave passages in the maze caves of the area (Chap. 1, Table 1.1), as well as the non-maze caves. Watertown, the location for several of the caves along the Black River has been the location of several Northeast Regional Organization (NRO) gatherings, with associated publications on the caves of the area (e.g. Fisher 1958; Zimmerman 1992). As there are a small number of studies relating to these caves only a few will be examined here with their relationship to glaciation, instead of an encyclopedic listing of caves and their descriptions. A particular focus will be on comparing the style of extensive caves in this region to those of the Helderberg Plateau.

7.2 Geologic History and Karst Setting

The lowlands of northern New York contain some of the oldest metamorphic and sedimentary rocks in the state. Outcrops include those of the Precambrian Grenville rocks (including the Grenville Marble, the karst forming unit of the Adirondacks Mountains; Chap. 9).

Fig. 7.3 Clints and grikes in the Ordovician Chaumont Limestone (Black River Group), west of Watertown, New York. **a** Flat limestone surface with joint-guided grikes at multiple angles, backpack in *circle* for scale. **b** Joint-guided grikes intersecting at right angles, backpack in background for scale. (Photos courtesy of T. Feeney)

As these are metamorphosed, there is a major structural component to them. The metamorphic rocks also include some igneous intrusions, as well as some overlying lavas (Cushing et al. 1910). Because of the metamorphic nature of these rocks, outcroppings of marble are not extensive and thus do not form many large cave systems. Several caves in this area do form in the Grenville Marble, such as the series of meander cutoff caves at Natural Bridge Cave (Mylroie and Mylroie 1990). As these caves are in the highly deformed Grenville Marble rather than the flat lying Cambro-Ordovician strata they will be examined with the Grenville Marble caves of the Adirondacks (Chap. 9), even though they lie on the low-relief outskirts of the Adirondack Mountains.

A band of metamorphic rocks connecting the Precambrian rocks of Ontario to the Adirondack Mountains, termed the Frontenac Axis provides a very erosionally resistant barrier, and acts as a drainage divide for the rivers and lakes of the region. Pre-glacially the Frontenac Axis acted as the drainage divide for Lake Ontario and the St. Lawrence River. Breeching of the Frontenac Axis by glaciation has allowed some rivers to divert to new, post-glacial drainage routes that are important in the determination the timing of cave origin (Fig. 7.1).

Above the Grenville rocks are several karst-forming Cambrian and Ordovician limestone units. These units include the Beekmantown Group, the Black River Group, and Trenton Group as well as the Chazy Limestone, Theresa Limestone and Little Falls Dolomite (Chap. 1, Fig. 1.2; Engel 2009). The Black River and Trenton groups are similar to the Helderberg Group that exists to the south (Chaps. 5 and 6) in that they are entirely carbonate units, with solubility differences between their formations. The more soluble units include the Leray Limestone and the Watertown Limestone (Cushing et al. 1910). Though these are the most soluble units, the other limestone units do form karst, including caves and limestone pavements (Fig. 7.3).

As noted above, the Cambro-Ordovician strata of this area has a low dip, typically less than 5° towards the southwest, with few other structures. The few structures that do exist in the Paleozoic rocks are joints that are dominantly oriented NW-SE, with a nearly orthogonal joint set oriented NE-SW (Fig. 7.4). In the soluble strata of this region the joints are clearly dissolutionally enlarged with limestone pavement on the surface (Fig. 7.3), and extensive joint controlled caves such as Glen Park Labyrinth in the subsurface (Fig. 7.2). Other important structures are monoclines that control the formation of waterfalls such as those of the Black River (Cushing et al. 1910).

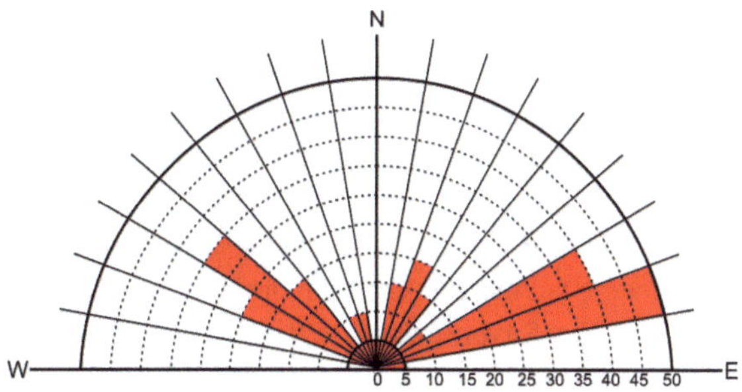

Fig. 7.4 Rose diagram of joint orientation and frequency in Paleozoic rocks of Jefferson County, New York. The joints here are oriented NW-SE, and NE-SW, which can be seen in the joint controlled maze caves along the Black River, such as Glen Park Labyrinth (Figs. 7.2 and 7.3). (Original data from Cushing et al. 1910)

This region has clear modifications by glaciation, including diversion of entire river systems, abundant limestone pavements, as well as abundant striations and erratics. The northern lowlands were among the last parts of the northeastern United States to be deglaciated, with complete withdrawal occurring ~13 ka BP (Ridge 2004).

7.3 Hydrologic History and Setting

The large scale hydrologic setting and history is very important to understanding the karst landscape evolution of the northern lowlands, perhaps greater than any other region in the northeastern United States. This region contains several larger rivers including the St. Lawrence River, the Mohawk River, the Black River, and the Perch River (Fig. 7.1). It also contains Lake Ontario, which is of particular importance, as lake effect meteorology greatly impacts the amount of precipitation recharging the local hydrology. The Mohawk River extends the furthest south, and locally controls cave base levels in the southern reaches of the northern lowlands. The Black and Perch rivers are additionally important, as many extensive maze caves exist within the walls of their channels, and in the case of the Perch River the entire river sinks beneath the surface near its outlet into Lake Ontario. The catchment area of these rivers seems to greatly impact the size of the caves formed, with the Black River having a catchment area of 5,000 km^2 (1900 miles^2), with a majority of the catchment upstream of the caves.

The history of the hydrology of this area is intimately tied to glaciation. During deglacial times Glacial Lake Iroquois existed in this region, as an ice bridge blocked the flow of water out the St. Lawrence River (Fig. 7.5). Deposits from this glacial lake can be seen near the maze caves of this region, including further east than the maze caves' locations. These deposits indicate that the areas where these caves are located were submerged until the draining of this lake at ~12.9 ka (Rayburn et al. 2011). The pattern of rivers in this area is strongly controlled by glaciation. In the vicinity of Lake Ontario the overall stream patterns are east to west, towards the lake. The Black River's entire course does not reflect this glacially influenced drainage pattern, and therefore it is interpreted that this river did exist in pre-glacial times.

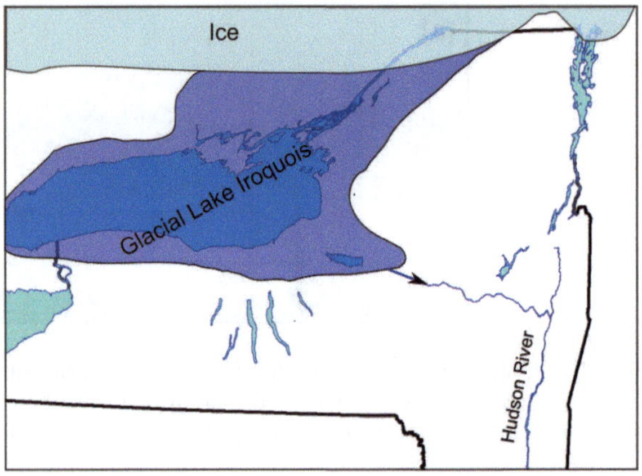

Fig. 7.5 Map of Glacial Lake Iroquois. This lake covered the northern New York Lowlands, extending to the Adirondack Mountains, and existed due to ice damming the St. Lawrence River. Lake drainage was routed through the Mohawk and Hudson rivers, with vastly larger discharges than seen at the current time. These large discharges cause the Mohawk and Hudson rivers today to be under fit in some locations (de Simone et al. 2008). The outlet that drained the lake is located near Rome, New York. (Modified from Wall and LaFleur 1995)

Though parts of the Black River follow a pre-glacial route, the river crosses the Frontenac Axis due to glacial diversion. As such, the river west of the Frontenac Axis exists in a post-glacial channel (Fig. 7.1), which is where the large maze caves are found.

7.4 Caves of the Northern New York Lowlands

The dearth of karst studies in Cambro-Ordovician rocks of the northern New York lowlands makes an exact discussion on the interaction of karst and glaciation difficult for this region beyond the study on maze caves by Cooper and Mylroie (2014). The cave systems of this region are mainly located around the border of Lake Ontario, and along the river systems draining into the lake. Other cave systems exist in the Mohawk River Valley, in some cases draining into that river. Many of the non-maze cave systems of the region are small, not well described scientifically, and lack glacial sediments, making their interpretations difficult. Here two caves with vertical components will be discussed (Schroeders Pants Cave and Mitchells Cave), as their passage style of formation may hint at their origins.

7.4.1 Schroeders Pants Cave

Schroeders Pants Cave is perhaps the most well-known cave of the northern New York Lowlands, located in Herkimer County. The fame of this cave does not come from its scientific contribution to karst, but instead from the death of a caver, James Mitchell, in 1965. The (failed) rescue effort and death made national news, and is chronicled in the *Northeastern Caver* following talks given at the National Speleological Society Convention about the death (e.g. Speece 2006). The remains of James Mitchell were recovered in 2006, and the cave was subsequently mapped in 2010 (Fig. 7.5; McKenna 2010). James Mitchell is commemorated by an annual award of the National Speleological Society for the best scientific paper presented by a person under the age of 25.

Schroeders Pants Cave had been well described in articles of the *NSS News* in the 1940s and within *Speleo Digest* in 1959 (Speece 2006). McKenna (2010) also provides a description of the cave, along with a map (Fig. 7.5). This cave is 434 ft (130 m) long, with a vertical extent of 134 ft (41 m) including a 100 ft (30 m) tall dome. The passage cross-sections indicate a vadose origin of the cave, with upper-level vadose passage channeling water to the dome (Fig. 7.6). These vadose canyons have widths of a maximum of 3 ft ($\sim$1 m). Dye tracing reveals flow through this cave is routed into East Creek (Speece 2006). The cave is decorated with flowstone and soda straws, some of which were damaged during rescue efforts (McKenna 2010).

Fig. 7.6 Map of Schroeders Pants Cave, Herkimer County, New York. This cave contains an upper, vadose level and a dome. Though the cave exhibits vertical development, this development is due to down-cutting in the vadose zone, and is not indicative of base level change that could require longer time periods to form. (Modified from McKenna 2010)

Caves with a vertical component may hint at their origin. Multiple levels of cave development sometimes indicate change in base levels that may be associated with glaciation. If a cave has multiple phreatic levels, perhaps with vadose entrenchment, this is a strong indication of base level change and may indicate that a cave has survived glaciation yet remained part of the new, post-glacial flow path. Passage widths and dome heights of vadose passage are poor indications of formation time for caves. Schroeders Pants Cave appears to be entirely vadose in origin with canyon shaped passages in the upper level, and a dome carved by vadose waters. The vadose, vertical nature of this cave does not give information about whether it is pre- or post-glacial in origin. The vadose nature of the cave, combined with recharge congruence with the post-glacial deranged drainage suggests post-glacial origins for this cave.

7.4.2 Mitchells Cave

Mitchells Cave is located on a limestone bluff overlooking the Mohawk River in Montgomery County, New York. As noted above, the cave has a vertical component that leads downward from the entrance; however at the bottom, 140 ft (43 m) below the entrance, a short system of roomy horizontal passages exists (Fig. 7.8). The cave captures only local diffuse input, and appears unconnected to the current hydrologic setting. The entrance and shaft series are placed near the edge of the escarpment leading to the Mohawk River below. The upper reaches of the cave show distinct phreatic and floodwater speleogens (Fig. 7.8b), not expected for a vadose shaft series. The Mohawk River in this area flows in a gorge through Ordovician limestones and dolomites. The gorge is post-glacial in origin (Isachsen et al. 2000). The cave may be pre-glacial, a relic from an earlier drainage pattern now entirely different, or the cave may have formed as part of the re-arrangement of hydrology as the modern Mohawk River established its current channel at the end of the last glaciation. The phreatic flow features found in the upper portion of the cave may indicate that the cave was a transfer route for glacial meltwater during the Lake Albany phase of the

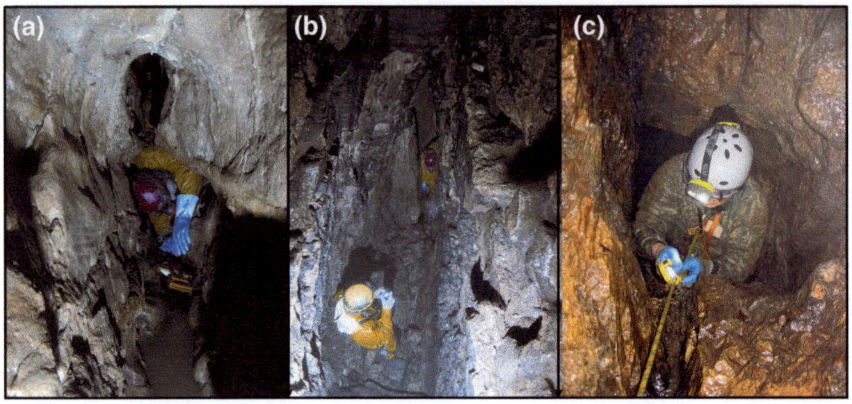

Fig. 7.7 Photographs from the interior of Schroeders Pants Cave, Herkimer County, New York. **a** The tight entrance series, which complicated rescue efforts in 1965. **b** The vertical descent at the end of the cave. **c** The top of the dome. (Photos courtesy of M. Chu)

last deglaciation, or could represent a sub-ice flow transfer system, as has been noted for some cave systems bordering fjords in Norway (Lauritzen and Skoglund 2013). Mitchells Cave is a good example of an intriguing but poorly studied cave, typical of the small, scattered caves found in the northern New York lowlands where lack of cave size, and broad dispersal of caves, has limited investigative focus by cave explorers, and subsequently, cave scientists. Comparing the plan view of Schroeders Pants Cave (Fig. 7.7) and Mitchells Cave (Fig. 7.8) shows them to look almost identical, but in profile view, one can easily see that Schroeders Pants Cave has its length in its upper vadose input passage, whereas Mitchells Cave has its length at depth in phreatic passage; a cautionary tale about interpreting caves from simple plan view maps.

7.5 Maze Caves of the Black and Perch Rivers

The caves mentioned above demonstrate cave styles that can be seen in other glaciated karst regions, and while they have some interesting historic and geologic aspects to them, they are dwarfed in size and geologic interest by the extensive maze caves along the Black River, and the Perch River. These maze caves include,

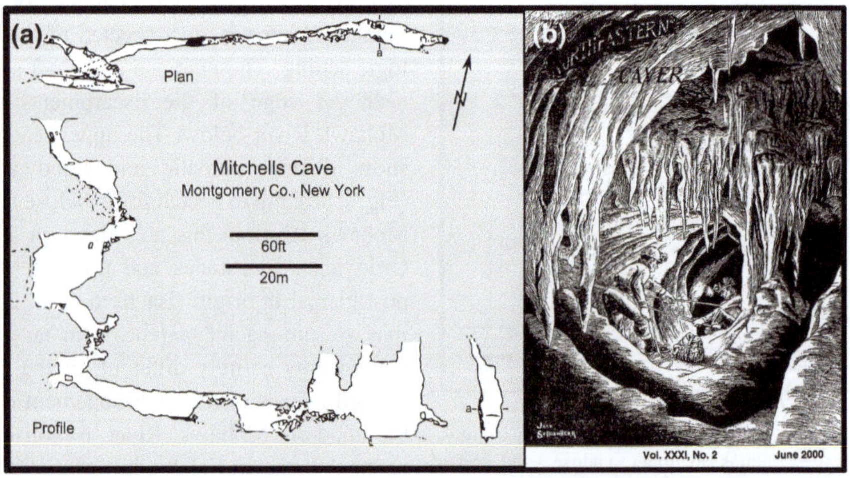

Fig. 7.8 Mitchells Cave, Montgomery County, New York. **a** Map of Mitchells Cave, showing vertical development leading to linear development at depth (from Evans et al. 1979). **b** Drawing by John Schoenherr (see Chap 3, Fig. 3.1 caption) displaying the top of the vertical tube leading to the cave's lower level. (from Porter 2000)

but are not limited to (in descending order of passage length): Glen Park Labyrinth, SCAG Cave, Three Falls Complex, Crayfish Water Labyrinth, Mangy Maze, and several others (Chap. 1, Table 1.1). Together the maze caves named here (with others adding additional length) contain over 12 km (7.5 miles) of cave passage (Engel 1996), greater than or equal to the entire passage length of all known individual dissolution caves in other regions of the northeastern US, or an entire region such as New England (~ 11 km; from Faulkner 2009, plus more recent discoveries; Chap. 8). Glen Park Labyrinth alone contains over 4 km (2.5 miles) of passage, placing it as the fourth longest cave in New York. Some areas of these maze caves include high cross-sectional passage areas, at up to 4.4 m^2 (47 ft^2) in Glen Park Labyrinth (Fig. 7.9; Cooper and Mylroie 2014). Parallel passages within the same cave greatly increase the amount of discharge they can handle. These caves tend to form at one level, though Glen Park Labyrinth contains several levels, with a lower maze and a still lower trunk phreatic passage (Carroll 1969a).

Several of these maze caves are relict, including Glen Park Labyrinth, the entrances of which are located 60 ft (18 m) above the active Black River. The maze caves along the Perch River such as SCAG Maze are actively forming, accepting the entire river in base flow. The drainage areas for these caves are much larger than those of any other region in the northeastern United States, with the Black River having a 5000 km^2 (1900 $miles^2$) drainage basin, most of which is upstream of the maze caves along the river. As such, the caves receive (or received, if relict) large discharges. For the actively forming maze caves, entrance into them is risky, and only during the summer at low flow can this entry be accomplished (Smith 1988). High flow rates are indicated in all these maze caves including the relict ones, in the form of scallops. These scallops are small (<2 cm; Fig. 7.9d) indicating water velocities at up to 3 m/s (Palmer 1975). Analysis of scallops, mean

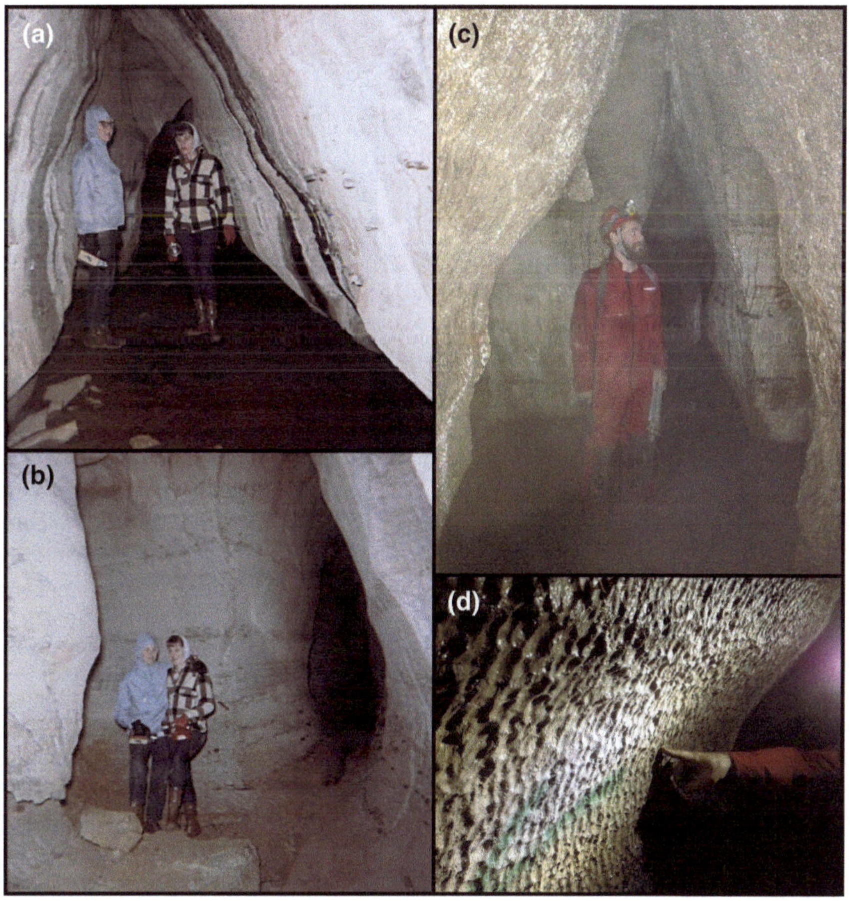

Fig. 7.9 Images of the Glen Park Labyrinth, Jefferson County, New York. **a**, **b** Pictures taken by Robert Carroll circa 1965, showing passage size (**a**) and passage junctions (**b**), courtesy of C. Porter. **c** Passage in 2013. **d** Scalloped wall in 2013 (with vandalism) indicating high-velocity water flow; when coupled with the passage size, the discharge was significant (see Cooper and Mylroie 2014). (Photos **c** and **d** courtesy A. Bartholomew)

cross-sectional areas, parallel passage, and potential, removed parallel passage for Glen Park Labyrinth indicates a top boundary condition of handling 159.3 m^3/s of discharge at maximum. A USGS gauging station upstream of Glen Park Labyrinth reports current statistical daily discharges with median values above 159.3 m^3/s (5626 ft^3/s) for 45 days per year (Cooper and Mylroie 2014), even with multiple impoundments towards the headwaters used to control discharges.

The maze caves of the Black and Perch rivers have impressive statistics in length, cross-sectional area, and capability to handle flow. The lengths (and some cross-sectional areas) of these caves rival those of the pre-glacial Helderberg Plateau systems. A question can then be asked: are the maze caves of the Black and Perch rivers pre-glacial, or post-glacial in origin? At first glance the statistics of these caves would make them appear pre-glacial; the length of Glen Park Labyrinth is less than 1-km shorter than Barrack Zourie Cave (Engel 1996), a clearly pre-glacial cave with post-glacial modification (Dumont 1995). Caves interpreted to be post-glacial, such as Westfall Spring Cave (Weremeichik and Mylroie 2014) of the Helderbergs, as well as passages within pre-glacial caves interpreted to be post-glacial overprints appear to have small cross-sectional areas. Arguments given by cavers of the northeastern US, following Palmer (1991), claim that while caves can form to traversable sizes in the time since deglaciation, larger passages should be pre-glacial or sub-glacial in origin to reach an appreciable cross-sectional area (e.g. Rubin et al. 1995).

The answer to this question is somewhat surprising; these maze caves are likely post-glacial in origin. Even the relict Glen Park Labyrinth, located 18 m above the current position of the Black River is post-glacial in origin. A simple look at the hydrologic history of the region answers the question in an intuitive way; as the Black River in the area west of Great Bend, New York is post-glacial, the caves controlled by the floodwaters of the river must be post-glacial in origin. The caves along the Perch River are active, and do not indicate a change in base level that would be associated with survival through glaciation. Additionally, though these caves exist within the footprint of Glacial Lake Iroquois, no lake sediment appears preserved in the caves. More complex analysis by Cooper and Mylroie (2014) using paleodischarge reconstruction, and a first principles approach using the wall-retreat rates generated when forming maze caves (~0.1 cm/year; Palmer 1999) demonstrates that Glen Park Labyrinth could form in the time since deglaciation (13.7 ka for this area; Ridge 2004) within the current discharge regime of the Black River. Additional modeling of paleodischarge supports formation in the time that Glacial Lake Iroquois existed over the cave location to the cave becoming relict above even the largest of modern-day flood stages of the Black River (Cooper and Mylroie 2014).

From the geologic and hydrologic history of this area a conceptual model can be formed explaining the development of these extensive maze caves. These caves likely initiated between a combination of deglacial and post-glacial mechanical enlargement of joints as a result of glacial unloading. The inundation by Glacial Lake Iroquois then provided high hydraulic gradients for quick breakthrough and initial conduit-full conditions year round. After the draining of Glacial Lake Iroquois the extensive drainage areas of the Black and Perch rivers generated large discharges allowing the caves to be in conduit-full conditions with floodwaters during a high percentage of the year. These conditions still persist along the Perch River, such as with SCAG Maze. Along the Black River headward migration of steep gradients (indicated by a waterfall upstream of Glen Park Labyrinth) has left some of the maze caves relict, at a position high above the current river level. This model allows the rapid formation of maze caves, and their quick placement into relict conditions.

The maze caves of the northern lowlands of New York therefore demonstrate that larger caves can be established in a post-glacial time period (and thus likely during the previous interglacial time periods), and the requirement needed to establish large caves in short periods of time is a high flux of aggressive waters. These floodwaters also can provide large amounts of mechanical erosion. The combination of aggressive waters generating large dissolutional wall-retreat rates and mechanical erosion forms extensive, voluminous caves rapidly. Floodwater maze caves therefore represent an interesting speleological phenomenon, and are similar to other caves that form at high rates in time limited regimes such as flank margin caves of carbonate islands that can form during the short interglacial time periods (e.g. Mylroie and Mylroie 2013b). Floodwater maze caves in glaciated terrains may also be time limited, as their shallow depths make them prone to removal by subsequent glaciation (Cooper and Mylroie 2014).

7.6 American History in the New York Lowland Caves

Houghs Cave, Lewis County, New York, is a small cave with a significant history (Fig. 7.10). The cave is stated to have been a stop on the Underground Railroad, used by escaped slaves fleeing to Canada to the immediate north, named after Horatio Gates Hough, Jr., an ardent abolitionist (Brown 2014). The cave site is identified by a New York State historical marker on Route 26, about a mile (1.6 km) south of the village of Martinsburg. The marker reads:

> ***HOUGH'S CAVE*** *HIDING PLACE OF RUNAWAY SLAVES BEFORE THE CIVIL WAR AND STATION ON THE UNDERGROUND RAILWAY TO CANADA.*

The configuration of the cave today as a low stream passage makes use of the cave by escaped slaves a bit of a problem, but as noted by Brown (2014), the cave appears to have been highly modified by dumping and agricultural land use in the immediate area, and the original cave passage size and shape cannot be determined effectively. The cave is developed in the Ordovician Lowville Limestone, and is shallow, with 344 ft (~100 m) of surveyed passage (Brown 2014). It is congruent with current post-glacial drainage, and is shallow, only a few meters below the surface (Fig. 7.10). These features suggest that the cave is post-glacial in origin.

Historical accounts of caves can be difficult to interpret, because the reliability of the initial observations may be questionable, and the modifications to the land and cave since those observations were made can add to the confusion of the modern observer. As noted in Chap. 5, John Cook (1906) provides a detailed and highly reliable map of portions of Howes Cave (now Howe Caverns) since quarried away, but other historical accounts, such as with Houghs Cave presented here, may be less reliable. Note the potential for error, as these two caves have similar sounding names, but have little else in common and are 130 km (80 miles) apart. Cave name changes can also generate errors; Gage Caverns in the Helderberg Plateau, has had four names in its post-1800 history: Balls Cave, Gebhardts Cave, Knoepfels Cave and Gage Caverns (Halliday 1976), and as presented in Chap. 5, has had its late 20th century name, Gage Caverns, changed back to an earlier name, Balls Cave, by its owner, the National Speleological Society. Such actions are certain to create problems for future cave scientists, as all recent scientific work in the literature refers to the cave as Gage Caverns. The *Northeastern Caver* runs a column in each issue called "Spelean Archives", where historical accounts of caves are presented, and

Fig. 7.10 Aerial view of Hough Cave, Lewis County, New York. The insurgence to resurgence distance is short, the cave is shallow as evidenced by the collapse entrances, and the cave is adjusted to the local, post-glacial drainage. These observations indicate the cave is post-glacial in origin (from Brown 2014)

discussed, providing insight into the long and varied history of northeastern US cave exploration and description.

7.7 Summary

The smaller caves and karst in the Cambro-Ordovician carbonates of the northern New York lowlands appear typical of those that appear similar in structure in the Helderberg Plateau (Chap. 5). Unlike those of the Helderberg Plateau, the caves of the northern lowlands are poorly studied (particularly the small caves). These small-in-length caves are a mixture of pre-, post-, and perhaps sub-glacial origin. The caves with vertical relief explored herein show two different styles of development: Schreoders Pants Cave as vadose versus Mitchells Cave as phreatic. These styles are hints towards their time origins and glacial controls. Small stream caves such as Houghs Cave meet the criteria of a post-glacial origin.

The striking difference between the caves of the Helderberg Plateau (Chap. 5) and the northern New York lowlands is the style and origins of extensive caves. In the Helderberg Plateau extensive caves of large cross-sectional areas are dendritic in plan, and pre-glacial in origin, surviving multiple glaciations with enlargement during interglacial times. These caves recapture the new, post-glacial deranged drainage and route flow through pre-glacially established flow routes. Relict caves and cave passage are likely pre-glacial in the Helderberg Plateau, though some examples exist of floodwater caves being post-glacial and containing relict passage (e.g. Onesquethaw Cave; Chap. 6). In contrast, the extensive caves of the northern New York lowlands are maze caves of post-glacial origin. These caves formed rapidly in deglacial and post-glacial times. Even relict maze caves may be post-glacial in origin. Unlike the pre-glacial systems of the Helderberg Plateau these post-glacial maze caves may be limited to the current interglacial, as their shallow depths make it likely they would be removed during the next glacial period. The pre-glacial systems of the Helderberg Plateau may be biased to preservation, as their long formation times allowed the establishment of deeper conduits and longer flow lengths. This comparison of extensive caves from these two areas of New York present a cautionary tale about use of simple metrics, such as mapped cave length, as a criteria for the timing of speleogenesis as pre- or post-glacial.

References

Brown B (2014) Houghs Cave trip reports. Northeast Caver 45 (3):107–110

Carroll RW (1969) A "moss-ball" linkup at Glen Park. Northeast Caver 1(3):104–106

Carroll RW (1972) Some recommended caves for the fall NRO. Northeast Caver 3(9):106–115

Cook JH (1906) Limestone caverns of eastern New York. In: Clarke JM (ed) Third report of the Director of the Science Division. New York State Museum, Albany, pp 32–51

Cooper MP, Mylroie JE (2014) Post-glacial speleogenesis: verification of a hypothetical model, and the origins of maze caves in glaciated terrains. Cave Karst Sci 41(2):84–95

Cushing HP, Fairchild HL, Ruedemann R, Smyth CH (1910) Geology of the Thousand Islands Region: Alexandria Bay, Cape Vincent, Clayton, Grindstone and Theresa Quadrangles. New York State Museum Bulletin No. 145–146, University of the State of New York

de Simone DJ, Wall GR, Miller NG (2008) Glacial geology of the northern Hudson through the southern Champlain lowlands. Guidebook to field trips for the 71st annual reunion, Northeastern Friends of the Pleistocene

Dumont KA (1995) Karst hydrology and geomorphology of Barrack Zourie cave system, Schoharie County, New York. Bulletin V of the New York Cave Survey

Engel TD (1996) New York State long cave list. Northeast Caver 27(1):11

Engel TD (2009) New York State. In: Palmer AN, Palmer MV (eds) Caves and Karst of the USA. National Speleological Society, Huntsville, pp 27–32

Evans J, Quick P, Sloane B (1979) An introduction to caves of the Northeast. National Speleological Society, Huntsville

Faulkner T (2009) Speleogenesis in New England marble caves. In: Proceedings of the Fifteenth International Speleological Congress, Kerrville, vol 2, pp 855–862

Feeney TP (1996) The role of grikes in limestone pavement formation in northern New York State, USA. In: Gines A, Fornos JJ (eds) Karren landforms. Universitat de les Illes Balears, Palma de Mallorca, pp 53–62

Fisher (1958) Caves of Watertown New York State. Northeastern Regional Organization, Berkley Heights, New Jersey, 54 p

Halliday WR (1976) Depths of the Earth. Harper and Row, New York

Isachsen YW, Landing E, Lauber JM et al (2000) Geology of New York: a simplified account, 2nd edn. New York State Museum, Albany

Lauritzen S-E, Skoglund RØ (2013) Glacier ice-contact speleogenesis in marble stripe karst. In: Schroder JF (ed) Treatise on geomorphology, vol 6, pp 363–396

McKenna P (2010) The Schroeders Pants Cave survey. Northeast Caver 41(2):47–48

Mylroie JE, Mylroie JR (1990) The consequences of meander incision into soluble rocks. Natl Speleol Soc Bull 52(1):33–44

Mylroie JE, Mylroie JR (2013b) Flank margin caves in carbonate islands and the effects of sea level. In: Schroder JF (ed) Treatise on geomorphology, vol 6, pp 351–362
Palmer AN (1975) The origin of maze caves. Natl Speleol Soc Bull 34(3):56–76
Palmer AN (1991) Origin and morphology of limestone caves. Geol Soc Am Bull 103:1–21
Palmer AN (1999) Patterns of dissolutional porosity in carbonate rocks. In: Palmer AN, Palmer MV, Sasowsky ID (eds) Karst modeling. Karst Waters Institute Special Publication 5, Charlestown, pp 71–79
Porter C (2000) (ed) Cover drawing. Northeast Caver 31(2)
Rayburn JA, Cronin TM, Franzi DA, Knuepfer PLK, Willard DA (2011) Timing and duration of North American glacial lake discharges and the Younger Dryas climate reversal. Quat Res 75:541–551
Ridge JC (2004) The Quaternary glaciation of western New England with correlations to surrounding areas. In: Ehlers J, Gibbard PL (eds) Quaternary glaciations—extent and chronology—part II: North America. Developments in quaternary sciences, vol 2. Elsevier, Amsterdam, pp 169–199
Rubin PA, Engel T, Nardacci M (1995) Geomorphology, paleoclimatology and land use considerations of a glaciated karst terrain; Albany County, New York. In: Garver JI, Smith JA (eds) Field trips for the 67th annual meeting of the New York State Geological Association, pp 81–107
Smith B (1988) A challenge in northern New York. Northeast Caver 19(1):23–26
Speece JH (2006) James G. Mitchell and the Schroeders Pants Cave tragedy. Northeast Caver 37(3):92–96
Wall GR, LaFleur RG (1995) The paleofluvial record of Glacial Lake Iroquois in the eastern Mohawk Valley, New York. In: Garver JI, Smith JA (eds) Field trips for the 67th annual meeting of the New York State Geological Association, pp 173–203
Weremeichik JM, Mylroie JE (2014) Glacial Lake Schoharie: An investigative study of glaciolacustrine lithofacies in caves, Helderberg Plateau, Central New York. J Cave Karst Stud 76(2):127–138
Zimmerman (ed) (1992) The lesser caves of Watertown. Northeast Regional Organization Publication, Watertown, New York

Caves and Karst of New England and Eastern New York

8

Abstract

Dissolutional cave development in the New England and eastern New York area is primarily in Cambro-Ordovician marbles that extend in a north-south band along the western boundary of Connecticut, Massachusetts, and Vermont, and similarly in the eastern-most portion of New York. Some Precambrian marbles, and some Ordovician carbonates are also found in this area. Maine has karst areas, but they are remote, low-lying, and poorly understood. Cave development appears to have been mostly post-glacial, with a few relict (pre-glacial) large caves (e.g. Aeolus Bat Cave, Morris Cave), and several large caves that are likely combination caves that while active in the current deranged hydrology, have passages inherited from pre-glacial times (e.g. Vermonster, Carthusian and Merlins caves). Joint activation by isostatic rebound following ice withdrawal, coupled with large glacial lake discharges, are argued to be the prime initiator of cave development. The caves that result are commonly shallow, and vulnerable to removal by the next glaciation, indicating a cyclic nature to cave development and destruction, avoided only by the larger, deeper or fortuitously placed cave systems. A general lack of glacial sediments in smaller caves is a good indicator that many caves are post-glacial, though some examples exist of pre-glacial caves containing sediment (e.g. Weybridge Cave). Cave science has been sparse and sporadic in this region, but a recent wave of papers, coupled with new discoveries of major caves, indicates that this region will be a greater participant in speleology in the future.

8.1 Introduction

New England contains 95,812 km^2 (22,985 miles^2) of land, only two-thirds that of New York's land area, and karst-bearing rocks are not abundant. Karst development is almost entirely in Cambro-Ordovician marbles, although some early Paleozoic limestones and dolomites exist in northwestern Vermont and to a greater extent in Maine (Chap. 1, Fig. 1.1b), and some Grenville-aged marble is present (Porter 2009). New York, east of the Hudson River, contains some early Paleozoic limestones and dolomites, and Cambro-Ordovician marbles, so for all practical purposes (except those political), it is part of the New England karst province and will be treated as such here (e.g. Merlins Cave, Columbia County, New York).

The majority of caves and karst terrains in New England are in marble, and therefore a metamorphic province. The non-carbonate metamorphic rocks

M.P. Cooper and J.E. Mylroie, *Glaciation and Speleogenesis*,
Cave and Karst Systems of the World, DOI 10.1007/978-3-319-16534-9_8,

adjacent to the marbles are predominantly schists. In this respect, New England differs significantly from the largest karst area of the region, the Helderberg Plateau of New York. First, the thrust tectonics and metamorphism has resulted in isolated pods and outcrops of marble in New England, therefore the catchment of most karst areas is limited and caves are consequently smaller. Second, in the Helderberg Plateau, the carbonate rocks are resistant and formed cliffs, and the clastic rocks formed lowlands; the large expanses of resistant carbonate-surfaced plateau area created many long cave systems. In New England, the metamorphosis of those clastic rocks into resistant schists, quartzites and phyllites, and the carbonates into less-resistant marbles, has inverted that topographic model, such that the marbles often form the valley floors near base level and the schists and quartzites form the mountains separating the valleys. In New England, glaciation enhanced this topographic disparity based on lithology. The majority of the largest caves result from outcrops surviving on valley sides, or where outcrop patterns are large to produce a greater catchment area, but some (e.g. Weybridge, Horse Farm Road, and Great Radium Springs caves) are in area-limited outcrops.

There has been little major scientific study of glaciation and karst development in New England; the vast majority of the literature about caves and karst comes from the caving community (e.g. the *Northeastern Caver*, as well as local grotto publications), with a secondary component from historical records. *New England's Buried Treasure*, a classic book by Perry (1946), is an interesting account of New England's caves and cave legends, as known at the time. The most thorough analysis has been for Vermont caves (Quick 1994, 2010, 2012). An interesting and ongoing paleoclimate study has very recently been done on Weybridge Cave in Vermont (Perzan et al. 2014). A significant paper on glaciation and karst in New England was done by a British citizen comparing New England karst to that of the Scandinavian Caledonides (Faulkner 2009). Use of basic karst investigative tools, such as dye tracing (e.g. Weybridge Cave, Quick 2010), or geophysics, has been minimal. The National Speleological Society's guidebook for the 2002 annual convention (Nardacci 2002) is the best single source of data and geology regarding caves in Maine, both dissolutional and pseudokarst caves. The relative paucity of data makes answering the basic question of glaciation and karst interaction difficult to address with specific examples, although Quick (2010) provides an excellent presentation about cave development in Vermont, demonstrating what conclusions can be made when an ample and detailed data base is available. Cave exploration has been on-going in the region for centuries. In the 1950s and 1960s, when cave exploration was leaping forward in the Helderberg Plateau, to be followed soon thereafter with a wave of scientific studies in the 1970s, things were more subdued in New England. Cavers who might have done more work in New England were drawn westward into the Helderberg Plateau with its extensive cave systems. In the last decade, however, a new wave of cave exploration, centered primarily in Vermont, Massachusetts and eastern New York, has made some major cave discoveries. This new wave of interest, if it follows the Helderberg Plateau pattern, may result in a series of scientific investigations in the near future that continue to address what is unique about this glaciated karst region. The work of Faulkner (2009), Quick (2010, 2012), and Perzan et al. (2014) may be a foreshadowing of such a cave research phase.

Karst rocks are not equably present in New England (Chap. 1, Fig. 1.1b). New Hampshire has almost no known soluble rocks. Rhode Island has only a few tiny karst areas (e.g. Moore 2007) and dissolution caves are small and rare. For Vermont, Massachusetts and Connecticut, the soluble rocks are concentrated in a north-south band along the western borders of those states. As noted earlier, this band spills over into extreme eastern New York State, almost entirely east of the Hudson River. Maine contains the largest surficial area of karst rocks, but many are in lowland swampy areas, and the areas are very remote and poorly visited, let alone studied (e.g. Island Cave, Marble Pond, Penobscot County, Maine); pseudokarst caves dominate the Maine cave literature (Hendrickson 2002, Chap. 4).

8.2 Caves, Karst and Glaciation in New England

Despite the lack of in-depth scientific study of glacial effects on cave and karst development in New England, some general statements can be made. Some caves are large and clearly not in agreement with the

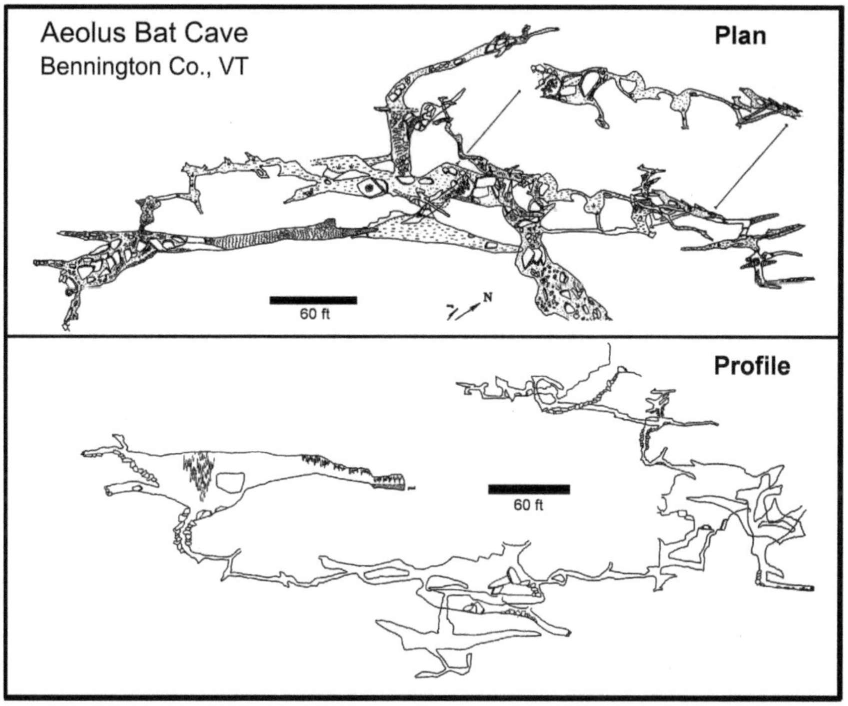

Fig. 8.1 Map of Aeolus Bat Cave, Bennington County, Vermont, showing plan and profile. (From Quick 2010)

current deranged landscape. They are no longer part of the main hydrologic cycle, and appear relict in both plan and position. The best example of such a senescent cave is Vermont's Aeolus Bat Cave in Bennington County, the largest cave in New England (Figs. 8.1 and 8.2; Chap. 1, Table 1.2). The cave has almost a kilometer of passage (~3100 ft), with a depth of 70 m (~230 ft), the second deepest karst cave in New England (only Purgatory Pit at 82 m or 269 ft is deeper; Higham 2013). The cave is located almost 500 m (~1600 ft) above the valley floor, and contains large chambers up to 70 m long and 12 m high (230 × 40 ft) with abundant calcite speleothems (Porter 2009). Both the large chamber size and the massive nature of some of the speleothems, coupled with the cave's placement high in the valley wall, indicate that the cave is almost certainly relict and pre-glacial in origin (Quick 2010). It is currently a major bat hibernaculum, which once had a bat population estimated at 300,000 (Davis and Hitchcock 1965), lowered to 23,000 bats in a 2003 census, and subsequently almost wiped out as a result of devastation from White Nose Syndrome or WNS (Porter 2009). The preferential use of this cave by bats (prior to WNS) may be a result of the cave's high and relatively dry position up on a valley wall, rare in New England. A full review of White Nose Syndrome in the northeastern United States can be found in Swezey and Garrity (2011).

In contrast, in northwestern Massachusetts, Coon Hollow Cave in Berkshire County (Fig. 8.3) is a small cave with small, active passages that seem adjusted to the current deranged drainage and landscape. It is one of many caves in this area, such as Eldon French Cave (Fig. 8.4), Coffin Cave and Convention Cave that are all small in passage size and in total length (Nardacci 1991), carrying streams in agreement with local drainage patterns. Twin Lakes Cave in Litchfield County, Connecticut (Fig. 8.5) is the longest cave in that state at 265 m (870 ft). It is well integrated with current drainage, displays phreatic/epiphreatic dissolutional features, but also contains abundant speleothems, and may represent a pre-glacial cave that exerted influence on the re-establishment of post-glacial drainage (see discussion below regarding "combination caves"). Some other small caves, such as Baker's Quarry Cave (Fig. 8.6), are remnants on valley sides well above base level, and may represent fragments of pre-glacial cave systems partitioned by glaciation.

Faulkner (2009) analyzed the existing cave database in New England (his Tables 1 and 2) to compare with similar data from central Scandinavia. He utilized

Fig. 8.2 Entrance images of Aeolus Bat Cave, Vermont. **a** Looking outward from the entrance at 760 m (2500 ft) elevation downward to the Vermont Valley *below*. **b** Inside the entrance chamber showing the spaciousness (for Vermont) of the passage. (Photos courtesy of J. Dunham)

153 marble caves totaling 9-km of passage that he characterized from the literature. Only five of the caves were actually visited in the field by him. Both western New England and central Scandinavia have cave development in Cambro-Ordovician marbles that have been glaciated. He states that the caves in New England match up well with the TDMO model (Top-Down, Middle-Outwards) developed for Scandinavia, which has four stages (Faulkner 2009, pp. 855–856):

1. *Rapid deglacial isostatic rebound causes seismicity and forms inception fractures to a maximum distance from the surface equal to one-eighth the depth of the local relief* (*tectonic inception*).
2. *Phreatic passages enlarge beneath flowing deglacial ice-dammed lakes (IDLs) over periods up to 2000 calendar years at relatively high wall-retreat rates, despite low temperatures and low* P_{CO_2} (*deglacial speleogenesis*).
3. *Mainly vadose passages entrench during interglacials* (*interglacial speleogenesis*).
4. *Glacial erosion removes whole caves or their upper and outer parts during the next glaciation* (*glacial removal*).

Ice dammed lakes (or IDLs) have not been a topic utilized to explain caves and their deposits west of the New England/Eastern New York area, however glacial lacustrine deposits are now recognized in the Schoharie County portion of the Helderberg Plateau (Weremeichik and Mylroie 2014, see also Chap. 5). Contrary to the Faulkner (2009) hypothesis regarding dissolutional aggressiveness, the argument in Schoharie County called for stagnant glacial lake waters to allow fine-grained limestone rock flour to settle out of the still water, without dissolutional removal. Neither sub-ice nor sub-lake dissolution has been reported from the Helderbergs. The explanation may be the degree and intensity of flow from ice dammed lakes through the New England marbles to promote speleogenesis. High discharge associated with deranged drainage and post-glacial lakes has been presented as a means to develop shallow maze caves by rapid dissolutional wall retreat in a variety of soluble lithologies in a number of New York settings (Cooper and Mylroie 2014). The Cooper and Mylroie (2014) study used first-principle dissolution kinetics to determine if the post-glacial time window was sufficient to allow cave development since regional deglaciation. Their work indicated that it was not only possible, but also likely. This work supports Faulkner's (2009) contention that the majority of New England dissolutional caves, especially the smaller ones, were post-glacial in origin, or at least highly modified by post-glacial water flow. Lauritzen and Skoglund (2013), who also studied cave development in Norway, have been critical of some aspects of the Faulkner (2009) model, based primarily on geochemical models. None the less, there seems to be agreement that in a post-glacial environment with glacial melt water or lake water in open-system conditions as regards P_{CO_2}, cave development as described can occur within the post-glacial time window, as also modeled by Cooper and Mylroie (2014).

Faulkner (2009, and references therein) classifies the caves in his study as one of three hydrological types: *relict caves* of pre-glacial origin, *mainly vadose caves* (or MV caves) of post-glacial origin, and *combination caves* which contain both relict pre-glacial passages and

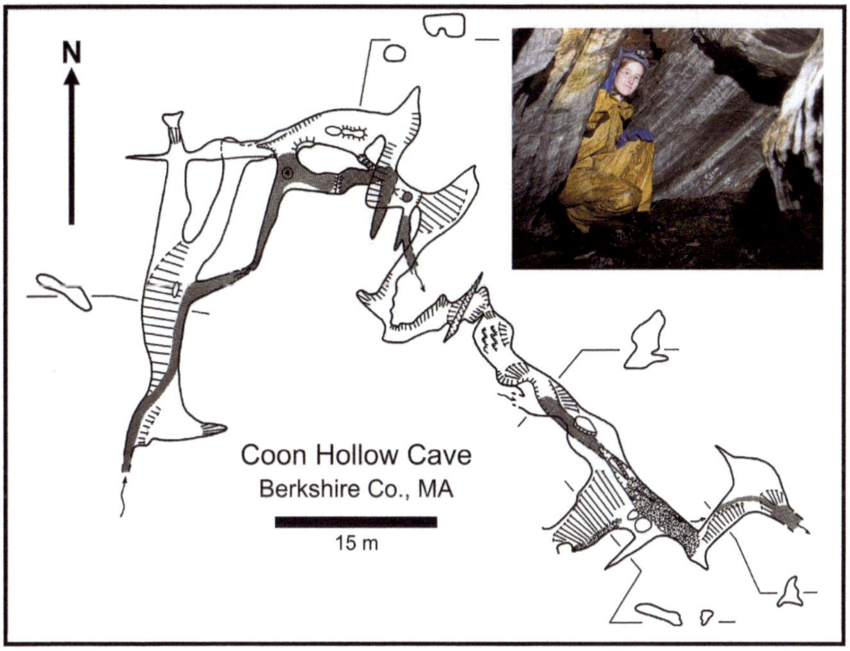

Fig. 8.3 Map of Coon Hollow Cave, Berkshire County, Massachusetts, from Evans et al. (1979). The cave has an active stream, and small passages. (Photo *insert* courtesy of J. Dunham)

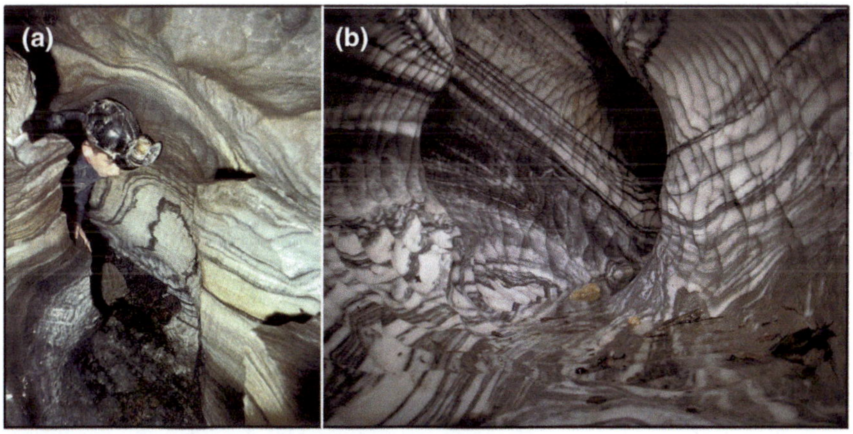

Fig. 8.4 Banded marble of Eldon French Cave, Berkshire County, Massachusetts. This cave is famous for the beauty of its marble walls. **a** Vadose canyon incised into the marble of Eldon French Cave (photo courtesy A. Palmer). **b** Clean-washed marble with abundant well-formed scallops; passage width in foreground approximately 1 m. (Photo courtesy J. Dunham)

active post-glacial, mainly vadose cave passages. In comparison with Scandinavian caves, Faulkner (2009) found that the New England caves were mostly in shallow dipping carbonate units (which he called *low angle karst*), with true stripe karst (where the carbonate rocks are thin bands dipping steeply) being uncommon. Despite the low dip of the cave-containing units, the mean vertical range of 9.3 m was quite similar to Scandinavia. This vertical range is believed by Faulkner (2009) to represent the one-eighth constraint (relative to valley depth) created by deglaciation unloading fracture depth. Vermonster and Carthusian caves in Vermont are in steeply-dipping marbles, which may bring their situation close to that of stripe karst.

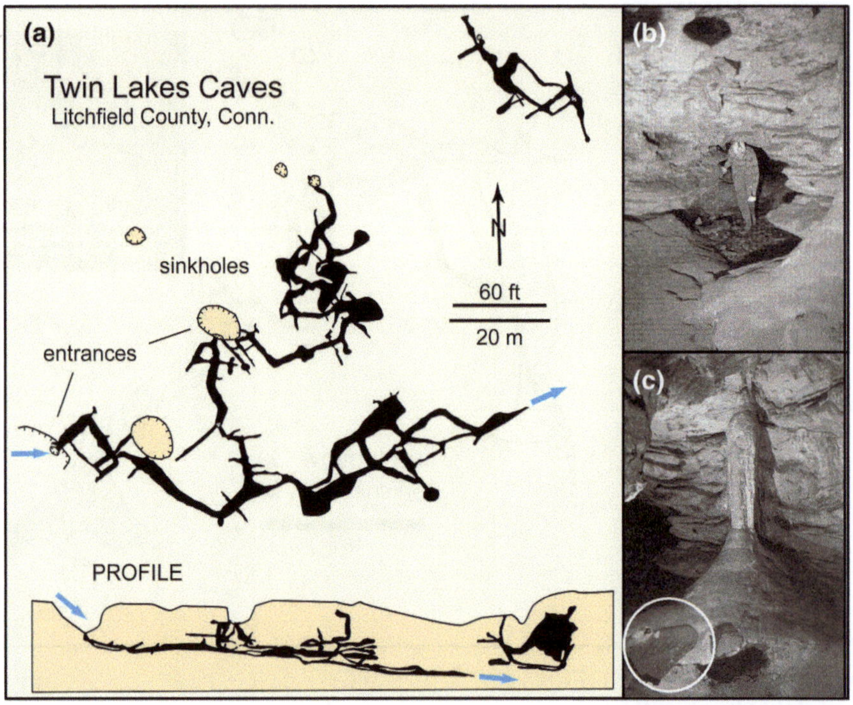

Fig. 8.5 Twin Lakes Cave, Litchfield County, Connecticut. **a** Map of Twin Lakes Cave, showing its linkage to current hydrology. (From Porter 2009). **b** Passage in Twin Lakes Cave, large for New England. **c** Bashful Lady formation, Twin Lakes Cave; this formation is large for New England and may be evidence of a pre-glacial cave origin. Circle shows a human leg bent at the knee, for scale. [Images **b** and **c** from Pease (2012)]

Fig. 8.6 Bakers Quarry Cave, Berkshire County, Massachusetts. (From Evans et al. 1979). An abandoned tubular cave with some vadose incision, it is de-coupled from current hydrology. (*Inset* photo courtesy of A. Palmer)

The vertical range of Vermonster is 189 ft (58 m), which greatly exceeds 22 ft (10 m), which when coupled with the cave's position high on the valley wall (2050 ft, 630 m), is in the range of the 1/8 rule. In other cases, while dye tracing data are lacking, the possible resurgences for many caves are at the valley floor, indicating other vertical flow routes of greater than 240 m (800 ft) in some cases (J. Dunham, pers. comm.).

8.3 The Vermont Example

The most detailed description of dissolutional caves in New England is that from the work of Peter Quick in Vermont (Quick 1994, 2010, 2012). He describes in a detailed summary the geologic and stratigraphic complexities of the Vermont Valley and the Champlain Valley to the north. The Taconic Orogeny was the main structural control, although the later Acadian Orogeny was influential, and the Triassic rifting of New England may have helped establish the current joint pattern, which Quick (2010) notes was later expressed by post-glacial isostatic rebound. The stacked thrust sheets of the Taconic Orogeny created a complex pattern of carbonate outcrops, including localized thickening of units, and the dispersal of carbonate rocks as small fault-plane slices. Quick (2010) adds that while the largest caves seem to be developed in the more highly calcitic units (e.g. Shelburne, Bascom and Chipman formations, Chap. 1, Fig. 1.3), these units are also favorably positioned up on valley walls where high hydraulic gradients and joint-gapping may have enhanced cave development, perhaps somewhat independent of carbonate purity. The loss of bedding planes to metamorphosis makes joints and faults critical to cave development.

Cave development in Vermont does not follow a regional water table, as in such complex and heterogeneous crystalline-rock geology a classic water table does not exist. Caves formed in response to very local hydrologic regimes, and floodwaters seem to have been a major factor (Quick 2010). Quick (2010), following Faulkner (2009), also implicates glacial lake floodwaters with early cave inception in Vermont. He also notes that phreatic passages are not common in Vermont, with only a few caves, such as Morris Cave (560 m or 1845 ft in length) showing abundant phreatic features, with later vadose overprint. The dominance of vadose passages reflects the location of the majority of the caves on valley walls with high hydraulic gradients, and isostatic-rebound opened joints, which promote downward vadose flow. Older, mostly phreatic caves are believed to have been completely or partially quarried away by more recent glaciations (Quick 2010).

Glaciation in Vermont was a significant factor in the present-day cave and karst landscape in that state. Ice sheet entry into the state probably began with the onset of the Quaternary, 2.6 million years ago (Quick 2010), and the ice of the last glaciation left the state between about 21,000 and 16,000 years ago. Ice thickness was 2500 m (8000 ft) over Vermont (Wright 2003). Lake Champlain underwent a series of changes, from ice-dammed fresh-water lake 15,000 years ago, to marine invasion to form the marine Champlain Sea 13,000 years ago, to fresh-water Lake Champlain as isostatic rebound separated the water body from its marine connection 11,000 years ago (Rayburn et al. 2011). Fresh-water lakes came and went as ice dams formed and disappeared, and as noted earlier, their flood-water discharge may have been the initiator of modern post-glacial caves in this area. The isostatic rebound, most rapid immediately after ice withdrawal, may have not only made joints open, but may have initiated earthquakes (Aylsworth et al. 2000), attesting to the strong forces acting on existing joints, fractures and faults in the region, which would have enabled those features to participate more completely in karst hydrology. In contrast, Lauritzen and Skoglund (2013) found little data in Scandinavia to support glacial rebound earthquakes as a major driver of fracture porosity.

Glaciation left behind large amounts of glacial sediments, both unsorted deposits such as tills and moraines, and sorted deposits such as outwash plains and lacustrine sediments. Quick (2010) notes that in Vermont, caves are almost never found where glacial sediments mask the carbonate rocks; only Morris Cave, opened by quarrying, exists below the glacial sediment limit in the valleys. This observation indicates that explorational bias is a factor; caves that cannot be found cannot be included in the cave database. That low elevation location may be why it shows more phreatic dissolutional features than most Vermont caves. In addition, Quick (2010) states that almost all caves in the Champlain Valley are found above the glacial sediment limit, but below the highest level of Lake Champlain and its predecessors. This observation strongly supports the glacial lake floodwater model for cave formation in the marble caves of New England, proposed by Faulkner (2009). Quick (2010) points out that the elevation of the caves as seen today is not the elevation at which they formed, because of continuing isostatic rebound.

Glacial sediments inside caves are not common in Vermont (Quick 2010), which again would support the contention that most of the caves are post-glacial in origin. Glacial sediment as clays are present in older, pre-glacial caves such as Morris Cave and Aeolus Bat Cave (Quick 2010), where they may represent fines that settled out under ice or lake cover. No analysis of these sediments has been done to determine if they are similar to those reported from the Helderberg Plateau by Weremeichik and Mylroie (2014). Bobcat Cave appears to contain glacial till that was extruded into the cave (Quick 2010); Bob's Birthday Cave has a similar deposit (J. Dunham, pers. comm.). One example of a cave containing dated sediments is Weybridge Cave, with Optically Stimulated Luminescence dates of sediments past that of the last interglacial, indicating that this cave is pre-glacial in origin (Perzan et al. 2014). Interestingly, Weybridge Cave (457 m or 1502 ft in length, one of Vermont's longer caves) is developed in the Belden Member of the Chipman Formation, which in this area is a limestone and not a marble (Quick 2010), as Vermont carbonates change from marble south of Rutland to limestone in the north.

8.4 Recent Discoveries

Recent work by cavers in the New England and eastern New York areas has made some major finds in the last decade. Merlins Cave in Columbia County, New York (Fig. 8.7), in Cambro-Ordovician marbles (Dunham 2013a), contains over 600 m (2000 ft) of passage as stacked levels in a primarily linear pattern, with a depth of 43 m (141 ft); the Big Room has a phreatic/epiphreatic dissolution signature overprinted by recent vadose flow. Vermonster Cave, western Vermont (Fig. 8.8; the specific location is suppressed for conservation reasons), of similar passage length and also in marble (Dunham 2014a), is somewhat linear and exhibits significant passage complexity (depth of 57 m or 186 ft). Most recently, Carthusian Cave was discovered (Dunham 2014b), similar in passage characteristics as Merlins and Vermonster, but smaller (as of 12 December 2104, 366 m, 1200 ft length with exploration on-going; depth is 44 m or 145 ft; J. Dunham, pers. comm.). These caves are currently active participants in the local subsurface hydrology, but they contain large passages and chambers, and it is a matter of question as to whether they are pre- or post-glacial in origin (e.g. Dunham 2013b); they could easily be both and fall into the *combination cave* category of Faulkner (2009). These recent discoveries indicate the importance of careful and aggressive fieldwork. While Aeolus Bat Cave has been known for a long time, recently discovered Merlins and Vermonster are the 2nd and 3rd longest dissolutional caves in New England/eastern New York (Higham 2013). Their size

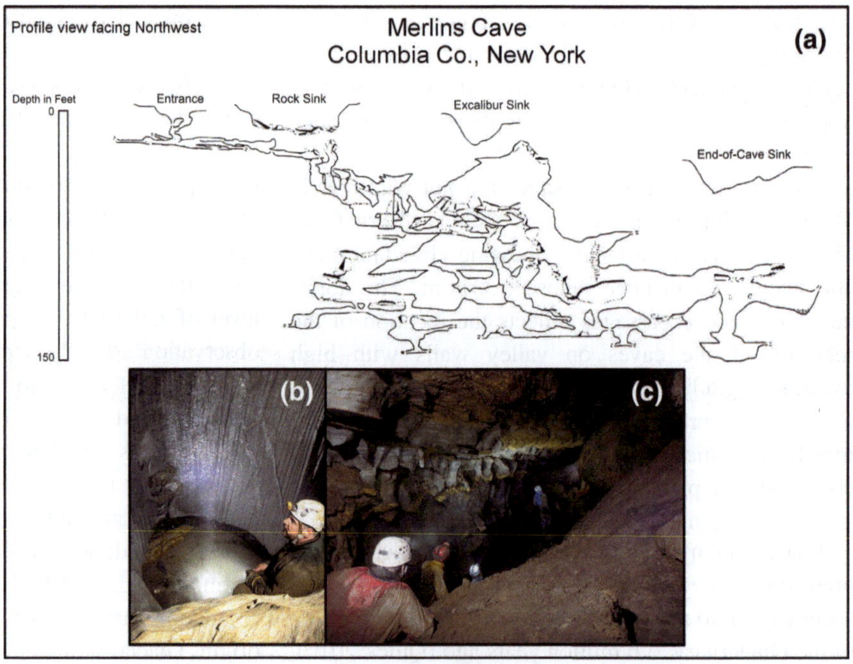

Fig. 8.7 Merlins Cave, Columbia County, New York. **a** Merlins Cave in profile (from Dunham 2013a), showing passage size and complexity. **b** Vadose shaft showing connection with current hydrology. **c** Large passage in Merlins Cave, perhaps indicating a pre-glacial origin. (Photos courtesy J. Dunham)

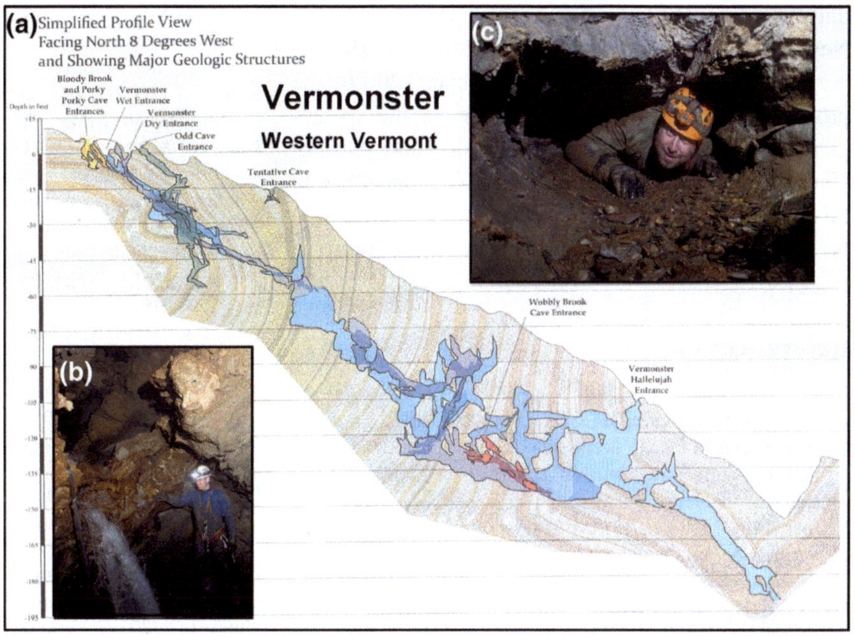

Fig. 8.8 Vermonster, western Vermont. **a** Profile map of Vermonster, showing depth, complexity, and large passage size (from Dunham 2014a). **b** Active vadose stream way in Vermonster, indicating coupling to current hydrology. **c** Small crawl passage in Vermonster; the passage size, the sculptured marble walls, and the glacially derived gravels are typical of New England marble caves. (Photos courtesy of J. Dunham)

re-opens the pre- versus post-glacial discussion; if pre-glacial, they argue for less cave removal by glaciation than previously thought.

8.5 Summary

The caves and karst of New England and eastern New York are developed primarily in marble, although Weybridge Cave and others in northwestern Vermont, are developed in limestone with dolomite interbeds (Quick 2010). The tectonic forces that created the marbles also resulted in marble outcrops being separated and partitioned such that large, continuous karst flow systems do not seem to be present. Large, pre-glacial caves exist, recognized as pre-glacial by a combination of position well above valley floors, decoupling from current hydrology, large chambers and passage size, and massive amounts of flowstone. Aeolus Bat Cave is a key example. Morris Cave, with its phreatic nature and burial under glacial sediments, would also seem to be relict and therefore pre-glacial. Other large caves, such as Merlins and Vermonster caves, are large and extensive, but are currently coupled to the modern hydrology. They may well be pre-glacial caves with a post-glacial overprint, the *combination caves* of Faulkner (2009). Glacial sediments are rare in the caves, but are more common in the few caves believed to be pre-glacial; by extension, those caves lacking glacial sediments are post-glacial in origin. The post-glacial caves are also recognized as they do not contain massive speleothems, are small in length and passage size, and are coupled to the modern hydrology. Some small caves, such as Baker Quarry Cave, are pre-glacial fragments.

The work of Faulkner (2009) demonstrated strong similarities between New England caves and caves from glaciated Scandinavia, also developed in Cambro-Ordovician marbles. His contention, supported by Quick (2010), is that deglaciation, which provides joint-gapping by isostatic rebound, and voluminous flooding from glacial lakes initiates speleogenesis at the end of each glacial cycle. The development of these caves in shallow positions makes them vulnerable to removal or segmentation by subsequent glaciations. Only a small percentage of caves survive from one interglacial to the next. The New York study on maze caves by Cooper and Mylroie (2014) supports these arguments of post-glacial origin and vulnerability to subsequent glaciations. The detailed work of Lauritzen and Skoglund (2013) from Norway reveals the type of data collection, analysis and modeling that are necessary to pin down the various arenas of cave development: pre-glacial versus englacial versus post-glacial. That level of precision has not yet been accomplished in New England.

The recent discoveries of large (for New England) caves suggests that explorational bias has hindered our understanding of glaciation and cave development in New England. As noted in Chap. 5 on the Helderbergs, and commented on by Faulkner (2009), the proximity of large population centers to the speleogenetic marbles of eastern New York and western New England has helped initiate a cave renaissance in this region, and it may drive more cave exploration to be followed up by more cave science in the future.

References

Aylsworth JM, Lawrence DE, Guetin K (2000) Did two massive earthquakes in the Holocene induce wide spread landsliding and near-surface deformation in part of the Ottawa Valley, Canada? Geology 28:903–906

Cooper MP, Mylroie JE (2014) Post-glacial speleogenesis: verification of a hypothetical model, and the origins of maze caves in glaciated terrains. Cave and Karst Sci 41(2):84–95

Davis WH, Hitchcock HB (1965) Biology and migration of the bat *myotis lucifugus* in New England. J Mammology 46:269–313

Dunham J (2013a) On the stability of big marble stream caves. Northeastern Cave 44(3):94–97

Dunham J (2013b) Surveying merlins cave. Northeastern Caver 44(1):6–9

Dunham J (2014a) Summing up the Vermonster. Northeastern Caver 45:42–53

Dunham J (2014b) Vermont's newest big cave. Northeastern Caver 45:71–77

Evans J, Quick P, Sloane B (1979) An introduction to caves of the Northeast. National Speleological Society, Huntsville

Faulkner T (2009) Speleogenesis in New England marble caves. In: Proceedings of the Fifteenth International Speleological Congress, Kerrville, vol 2, pp 855–862

Hendrickson E (2002) Caves of Maine. In: Nardacci MI (ed) A guide to the geology and caves of the Acadian Coast. National Speleological Society, Huntsville, pp 8–21

Higham S (2013) Northeast long caves, northeast deep caves. Northeastern Caver 44(2):44–45

Lauritzen S-E, Skoglund RØ (2013) Glacier ice-contact speleogenesis in marble stripe karst. In: Schroder JF (ed) Treatise on geomorphology, vol 6. pp 363–396

Moore J (2007) Early cave compilations of Rhode Island. Northeastern Caver 38(1):14–16

Nardacci M (ed) (1991) Guide to the caves and karst of the northeast. National Speleological Society, Huntsville

Nardacci M (2002) A guide to the geology and caves of the Acadian Coast. National Speleological Society 2002 Convention Guidebook. Prints Charming, Boothbay Maine, p 91

Pease B (2012) Brian Pease photos. Northeastern Caver 43 (3):94–95

Perry CW (1946) New England's buried treasure. Stephen Daye, Brattleboro

Perzan Z, Munroe JS, Amidon WH (2014) Investigation of last interglacial sediment in Weybridge Cave, Vermont, USA. In: Geological Society of America, Abstracts with Programs, vol 46(6). p 349

Porter C (2009) New England. In: Palmer AN, Palmer MV (eds) Caves and karst of the USA. National Speleological Society, Huntsville, pp 24–26

Quick PG (1994) Vermont caves: a geologic and historical guide. Paleoflow Press, Ypsilanti

Quick PG (2010) Vermont caves: a geological and historical guide. National Speleological Society, Huntsville

Quick PG (2012) Vermont caves: a geologic and historical guide. Paleoflow Press, Jaffrey

Rayburn JA, Cronin TM, Franzi DA, Knuepfer PLK, Willard DA (2011) Timing and duration of North American glacial lake discharges and the Younger Dryas climate reversal. Quaternary Res 75:541–551

Swezey CS, Garrity CP (2011) Geographical and geological data from caves and mines infected with white-nose syndrome (WNS) before September 2009 in the eastern United States. J Cave Karst Stud 73(3):125–157

Weremeichik JM, Mylroie JE (2014) Glacial Lake Schoharie: an investigative study of glaciolacustrine lithofacies in caves, Helderberg Plateau, Central New York. J Cave Karst Stud 76 (2):127–138

Wright S (2003) Glacial geology of the Burlington and Colchester 7.5 minute quadrangles, northern Vermont. Vermont Geological Survey, pp 1–12

9 Caves and Karst of the Adirondacks

Abstract

The Adirondack Mountains are the youngest mountain range in the northeastern US and are still doming, although they are made of the oldest rocks found in that region. The 1.1 Ga-old Grenville Marble hosts the dissolutional caves, which appear to be primarily post-glacial in origin, although the complexity of some caves (Crane Mountain), or large passage size (Burroughs Cave, Natural Bridge Cave, Natural Stone Bridge and Caves), suggest a pre-glacial origin. These large caves are all coupled to the current deranged hydrology, which makes them potential combination caves. Maze caves have formed (Big Loop Cave and X Cave), and follow the pattern for maze caves examined in the rest of New York: shallow, single level, in concert with the current deranged drainage pattern, and in a past or current high-discharge setting. These maze caves are not multi-level like those in similar lithologic settings, as in the stripe karst of Norway, as dip angles and hydraulic gradients are not as extreme. The meander cutoff cave situation present at Natural Bridge Cave mimics that found in flat-lying Paleozoic limestones of Kentucky. The development of meander cutoff caves and maze caves in glaciated 1.1 Ga-old marbles that are identical to patterns seen in undisturbed telogenetic limestones demonstrates how the overall hydrologic regime controls speleogenesis. As in other areas, scientific study of these caves has been minimal and many opportunities for research are present.

9.1 Introduction

The Adirondack Mountains (also termed Adirondacks) of New York (Fig. 9.1) are located in the northeastern portion of the state (Fig. 2.1), bounded by Vermont to the east, the northern New York lowlands to the north and west, and the Taconic Mountains and Mohawk and Hudson River Valleys to the south. These mountains are the youngest mountains in New York, and are still uplifting. Despite the young age of the mountains, the rocks that compose the uplifting dome are the oldest of the entire northeastern United States. Rocks of the Adirondack Mountains are metamorphosed, and include schist, gneiss, meta-anorthosite, and marble over 1 billion years in age (Isachsen et al. 2000).

Several caves within the Adirondacks appear on the northeastern United States long caves list (Chap. 1, Table 1.2), giving the appearance of large, extensive cave systems; these caves however are talus caves or crevice caves, and are not formed by dissolution (Chap. 4). Both by areal extent of exposed carbonates,

M.P. Cooper and J.E. Mylroie, *Glaciation and Speleogenesis*,
Cave and Karst Systems of the World, DOI 10.1007/978-3-319-16534-9_9,

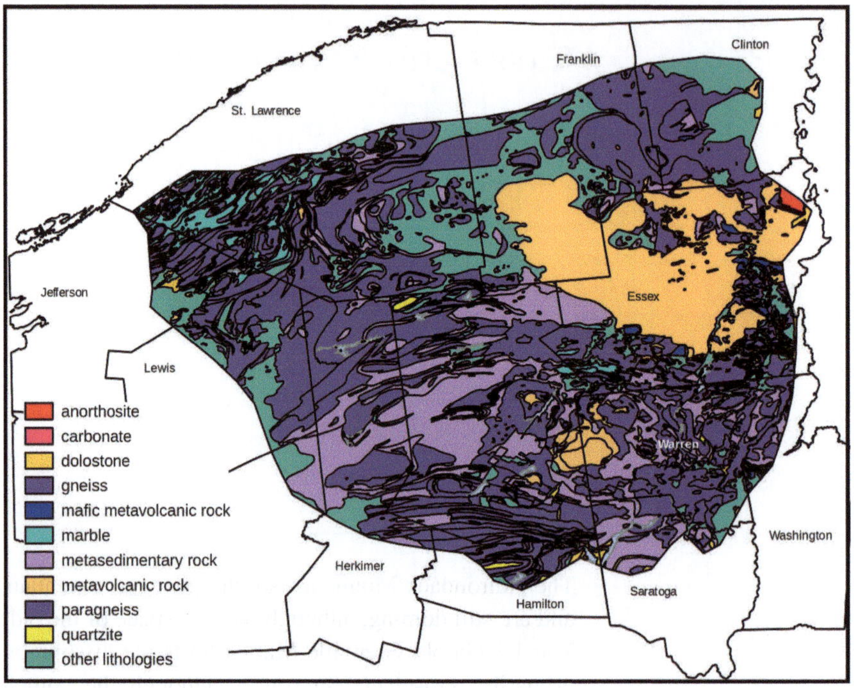

Fig. 9.1 Geologic map of the Adirondack Mountains. The geology of this region consists of Grenville aged (1.1 Ga) metamorphic rocks, including schist and marble. This region also contains an uncommon rock type, meta-anorthosite. Extensive non-dissolution caves form in these metamorphic rocks due to oversteepened slopes of escarpments and U-shaped valleys (Chap. 4). Limited outcroppings of marble tend to limit extensive dissolutional cave development

and by total known dissolutional cave length this karst region is the smallest in the northeastern United States. While total extent of cave development is small, this area is fairly important to discussions of karst. The caves seen here are formed in highly deformed, 1.1 Ga marbles, yet can show similar developmental styles to those formed in flat lying, well-studied karst regions such as Kentucky (Mylroie and Mylroie 1990).

A direct comparison of the Adirondack karst to other, well-studied karst areas is difficult to formulate, as this region is one of the few areas where Precambrian marbles outcrop in a glaciated terrain. A direct analogous terrain is the Canadian Shield, of which the Adirondack Mountains are a southward extension. Faulkner (2009) offers a comparison of the Adirondack karst to that of the Norwegian and Scottish Caledonides (and in extension, New England; Chap. 8), and applies his top-down-middle-outward (TDMO) model to selected caves here.

Like the northern New York lowlands and New England, the caves and karst of the Adirondack Mountains have received little attention in the scientific literature, as this area contains few extensive caves, and is among the most remote locations in the northeastern United States (except for the interior of Maine). Faulkner (2009), and Cooper and Mylroie (2014) both discuss caves in the Adirondacks. Mylroie and Mylroie (1990) also discuss caves located within the Grenville Marble near Natural Bridge, New York. Despite the relatively few caves in such a large area as the Adirondacks, there are two show cave operations using natural bridges in the Grenville Marble. The above-mentioned Natural Bridge is in Jefferson County, New York, on the west side of the Adirondacks just east of Watertown, New York, and then Natural Stone Bridge and Caves is on the east side of the Adirondacks (Engel 2005), just west of I-87 in Pottersville, Warren County, New York. Both systems will be discussed in this chapter, and it is important not to confuse the two. Again, like the other less-studied regions, descriptions do exist of caves in the Adirondacks in *The Northeastern Caver*, and in regional guidebooks such as Evans et al. (1979), Nardacci (1991), and Quick (2010). No extensive studies such as those in the Helderberg Plateau (Chap. 5), or in the Hudson Valley Fold-Thrust Belt (Chap. 6) have been performed describing pre-glacial cave systems that may be indicated by glacial sediments, multi-level phreatic systems, or truncation along valley sides. As such, the approach here will be to examine several caves that have been studied, rather than an exhaustive list of caves in the region.

9.2 Geologic and Hydrologic Setting

The Adirondack Mountains are a currently uplifting dome. This uplift began in the Neogene, making these mountains among the youngest in the world, despite being composed of 1.1–1.3 Ga metamorphic rocks (Isachsen et al. 2000). The metamorphic rocks that outcrop here are the common metamorphic rocks: phyllites, schists, gneisses, quartzites, and marbles, as well as meta-anorthosite. The uplift of these metamorphic rocks makes the Adirondacks the area of the highest elevation and highest relief in New York State. As mentioned in Chap. 8, marbles tend to form lowlands in metamorphic provinces, as they are less resistant than other metamorphosed rocks, and that is the case in the Adirondacks. The metamorphism and high amounts of deformation of these Precambrian rocks makes it difficult to establish a clear set of formations, and thus these units are broadly termed the Grenville metamorphic province (Isachsen et al. 2000).

The marbles of the Adirondack Mountains (Fig. 9.2) are commonly found sandwiched between schist units. This relationship is similar to that of the stripe karst in Norway, where karst developed in highly dipping beds form network mazes (e.g. Lauritzen and Skoglund 2013). The marbles in the stripe karst of Norway have limited surface exposure due to their highly dipping nature, and therefore produce "stripes" of marble outcrops. As the surface exposure of the marble is small, and the network mazes are formed more deeply than is typical of floodwater mazes in low dip strata, it may be possible for these to survive subsequent glaciations. Despite similar stratigraphy, the schist and marble units of the Adirondacks do not form classic stripe karst due to low dip angles (<30°; Faulkner 2009).

The surficial geology, topography, and drainage of the Adirondacks have clearly been shaped by glaciation, with glaciers having retreated entirely from the area only ~13.9 ka (Ridge 2004). Unlike within the

Fig. 9.2 Outcrops of Grenville Marble in the Adirondack Mountains of New York. **a** Outcrop along Route 8, Hamilton County, New York (south-central Adirondacks). **b** Distinctive marble pattern, Natural Stone Bridge and Caves, Warren County, New York (east side of the Adirondacks); this marble contains graphite which could be considered "Precambrian coal" derived from stromatolites. **c** Coarsely crystalline marble weathering out as a marble "grus", Jefferson County, New York (west side of the Adirondacks). **d** Distinctive marble pattern, Natural Bridge Cave, Jefferson County, New York (west side of the Adirondacks). Lens cap in c and d is 6 cm in diameter for scale

other karst regions of New York the Adirondack Mountains contain broad, U-shaped valleys (Fig. 9.3) as well as hanging valleys with waterfalls, typical of alpine glacial erosion. Glacial erosion left many depressions within the Adirondacks, as well as till and other glacial sediments, which have altered the surface drainage; waters pond to become lakes, and if the drain is inefficient, swamps. These lakes are connected by small streams guided by depositional landforms, and are part of a highly deranged drainage system (e.g. Fig. 9.4). The Adirondacks also play a large part in the overall hydrology of New York State, with the head-waters of the Hudson River, as well as the Black River, being located in this region.

9.3 Big Loop Cave

Big Loop Cave is one of the few caves that have been featured in a scientific study within the Adirondack Mountains (Cooper and Mylroie 2014), and is located in Essex County, New York on the Dutton Mountain 7.5 × 15-min quadrangle. This cave has also been well described by Engel (1989), with attention paid to geologic structure, hydrology, and topography. Big Loop Cave is an anastomotic maze cave (Fig. 9.5), with passages dipping at 15°, following the local dip of the marble that hosts the cave (Engel 1989). This maze consists of a large closed loop, with several smaller closed loops. The shapes of these passages are

Fig. 9.3 Photograph of broad, U-shaped valleys in the Adirondack Mountains, currently occupied by rivers. Unlike elsewhere in New York State, the Adirondacks contain many U-shaped valleys. Photograph taken on ascent to Eagle Cave in Hamilton County, New York (Photo courtesy M. Chu)

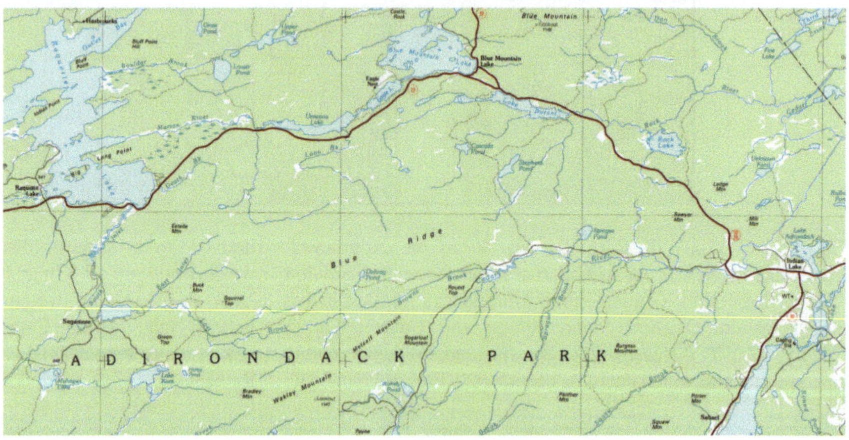

Fig. 9.4 Section of the Raquette Lake USGS topographic quadrangle, located within the Adirondack Mountains. The drainage shown here is highly deranged, with lakes and swamps located in glacial depressions, connected by small streams

fissures and tubes, with some vadose entrenchment into the floor (Fig. 9.5c). Careful surface surveying reveals that the cave is entirely within a maximum of 15 ft (~5 m) thick marble.

This cave is fed by a sinking stream that clearly fits within the deranged drainage condition of the Adirondacks (Fig. 9.5b). In the down-dip direction (approximately northwest) Sump Passage contains water year round. The passage Snorkel Crawl contains water year round, but is not sumped. During high flows the entire cave becomes conduit-full (even the passage dubbed "Dry Channel"), as indicated by large-sized organic debris found throughout the cave. Flooding has also emplaced rounded cobbles into the cave.

Big Loop Cave is likely post-glacial in origin; this cave fits within the deranged drainage, does not contain any autogenically emplaced glacial sediments, and can have formed to its current passage dimensions based on the duration it spends in conduit-full condition per year in the time since deglaciation (Cooper and Mylroie 2014). The shallow depth of the cave (5 m or 15 ft depth at maximum) also makes it seem that this cave is post-glacial, as it would not have likely survived erosion from the previous glaciation. This maze cave therefore also fits the hypothesis of Cooper and Mylroie (2014) that maze caves in glaciated terrains are post-glacial in origin. Big Loop Cave therefore provides an interesting end-member condition to this hypothesis; even caves within marble can form rapidly in the time since deglaciation. Therefore, it is the aggressiveness and mechanical transport properties of floodwaters that allow rapid formation, irrespective of whether the cave is in limestone, or marble. The case of Big Loop Cave is not similar to the stripe karst of Norway, even though it exists in a similar lithologic setup of marble sandwiched between two schist units, as the controls on stripe karst development are high dip angles and speleogenesis aided and guided by lithologic contacts, in addition to the steep hydraulic gradients associated with deep fjords.

Big Loop Cave also does not exactly follow Faulkner's (2008) TDMO model, as passages within the cave are predominantly phreatic, with some vadose modification, rather than mostly vadose. Though this is the case, the tube and fissure passages of the cave are epiphreatic, and are guided by flooding rather than constantly existing within the phreatic zone. The speleogenesis of Big Loop Cave can be approximated by the TDMO model, with high gradients and mechanical enlargement of chaotic fractures within the marble initiating speleogenesis, followed by conduit-full conditions nearly year round, and then epiphreatic development during intense flood events such as during spring snowmelt as the drainage basin matures. Like the other maze caves examined by Cooper and

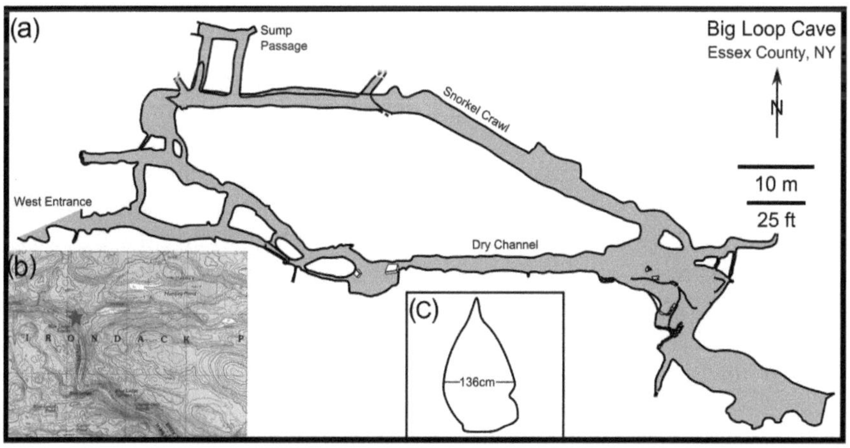

Fig. 9.5 Map of Big Loop Cave, Essex County, New York. **a** Map of the cave, showing anastomotic maze geometry with multiple closed loops. The complex geometry is a result of the chaotic nature of joints in marble, and floodwaters. (Map adapted from Engel 1989). **b** Location map of Big Loop Cave, indicated by a star on the USGS Dutton Mountain 7.5 min quadrangle, the cave fits well into the deranged drainage. **c** A typical cross-section of passage in Big Loop Cave, taken in the Dry Channel passage. The dimensions of these passages can be established in a post-glacial time window (Cross-section from Cooper and Mylroie 2014)

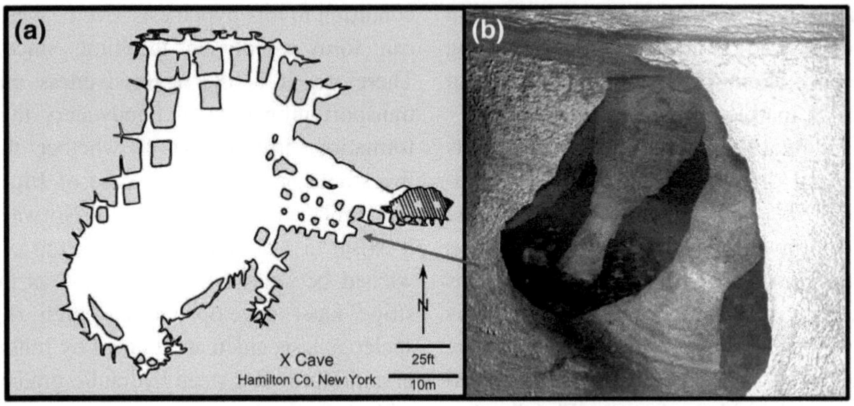

Fig. 9.6 X Cave, Hamilton County, New York. **a** Map of X Cave, showing a maze perimeter surrounding a large room, which is basically a consolidated maze (adapted from Quick 2010). This large room, like large rooms in other maze caves is a result of breakdown as dissolved pillars can no longer support ceiling spans. **b** Remnant dissolutional pillars near the wall of the cave, scene about 1 m across. Photo taken by Bob Carrol in 1970, courtesy of C. Porter

Mylroie (2014) this cave is shallow, and is likely to be removed by future glaciations.

The case of Big Loop Cave may also give hints towards speleogenesis of other maze caves in the marbles of the Adirondacks. As noted above and by Faulkner (2009), relatively low dips in the Adirondack Mountains do not form stripe karst, despite the similar lithologic patterns. While not examined by either Faulkner (2009), or Cooper and Mylroie (2014), a nearby network maze cave, X Cave (Fig. 9.6) may have a similar speleogenetic history, also following the modified TDMO model.

9.4 The TDMO Model in the Adirondack Mountains

While Big Loop Cave and perhaps other maze caves in the Adirondack Mountains do not completely align with Faulker's (2008) TDMO model, other caves examined, along with caves in New England (Chap. 8) do line up with this model. Caves from the Adirondacks that appear within Faulkner's (2009) paper on New England, the Adirondacks, and Newfoundland are: Crane Mountain Cave (Fig. 9.7), Browns Cave, Rusty Stove Cave, and Burroughs Cave (Fig. 9.8). Each of these caves fit well within the TDMO model, including the maximum one-eighth depth of fractures relative to their local glacial valleys (Faulkner 2009).

As the caves studied by Faulkner (2009) appear to fit into the TDMO model, it can be interpreted that these caves, like the maze caves discussed above, are post-glacial in origin. There are some problems with this post-glacial interpretation, as some caves, such as Burroughs, have very large chambers that may have required more time to develop than the ~13,000 years since Adirondack de-glaciation. This issue will be revisited later.

9.5 Meander Cutoff Caves in the Grenville Marble

The natural bridge and associated caves in the Grenville Marble at Natural Bridge, Jefferson County, New York, was used as an example of meander cutoff cave development by Mylroie and Mylroie (1990). That paper examined the situation where meandering streams incise into soluble rocks. Dissolution through the meander necks to produce meander cutoff caves has impacts both upstream and downstream (Fig. 9.9). This cutoff cave formation is driven by stream water being able to flow through the cutoff cave to the same downstream destination as the meander loop water over a shorter distance, hence a steeper gradient is produced, which drives speleogenesis. The presentation of two examples from flat-lying Mississippian limestones in western Kentucky was used to establish the primary factors involved in meander cutoff cave development. These two sites were then compared to the Natural Bridge setting, to see if the model would hold up when viewed in a much more complex area.

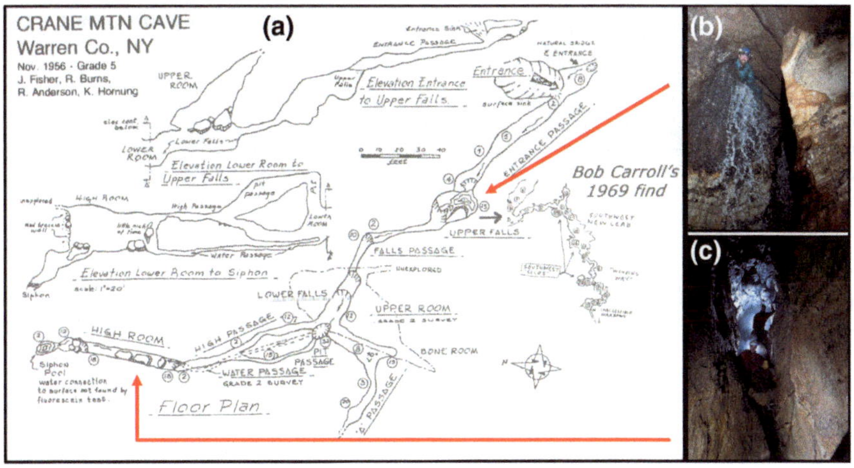

Fig. 9.7 Crane Mountain Cave, Warren County, New York. **a** Map of the cave, which displays multiple levels, a large, high room, and an upper level sediment choke, all of which might indicate a pre-glacial origin. Scale bar is 40 feet (12 m) (from Nardacci 1991). **b** Waterfall in the entrance series (Photo courtesy of M. Chu). **c** Upper level phreatic tube, downstream end of the cave (Photo courtesy J. Dunham)

The Grenville Marble of Natural Bridge is 1.1 Ga, and has been recently glaciated, a very different setting from the simple "type" examples of Kentucky. The meander cutoff cave pattern seen at Natural Bridge followed the pattern established for the simpler Kentucky setting, indicating the strong influence of meander incision on speleogenesis overall (Fig. 9.10).

The Mylroie and Mylroie (1990) study did not directly address the issue of glaciation and cave development at Natural Bridge; it was sufficient to demonstrate that the area had been glaciated, and that very old rock undergoing very recent geologic trauma behaved as predicted after stream meander incision by the Indian River (Fig. 9.10). Carroll (1969) had done the first documentation of these caves, and Mylroie and Mylroie (1990) followed his nomenclature for the caves (Fig. 9.11). The study by Mylroie and Mylroie (1990) built on earlier work done for a cave resource

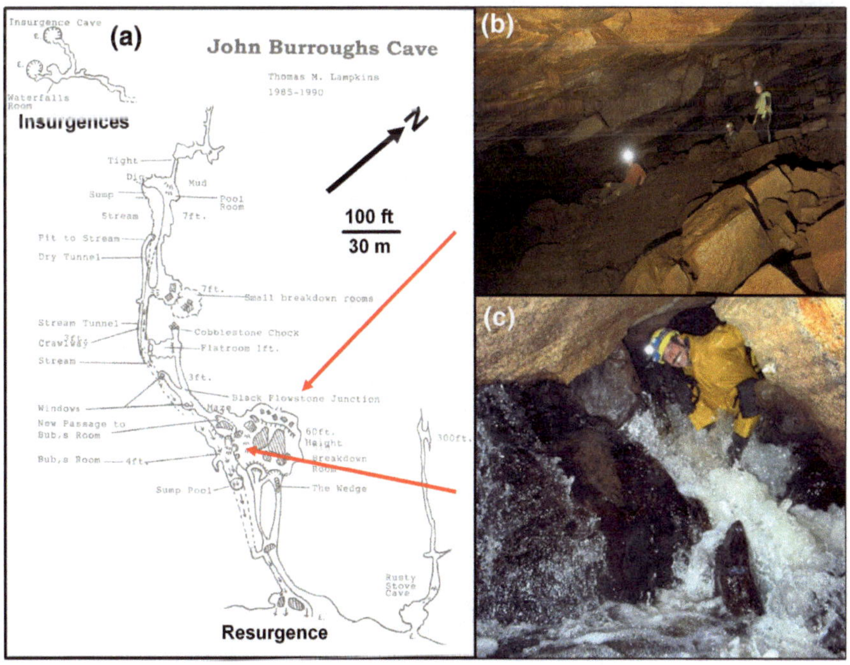

Fig. 9.8 Burroughs Cave, Essex County, New York. **a** Map of Burroughs Cave and adjacent caves; the cave is well integrated into the post glacial drainage, but contains a room over 30 m wide and 18 m high (100 ft wide and 60 ft high) (from Nardacci 1991). **b** Image of the large collapse chamber. **c** Image of the vadose stream that flows beneath the breakdown chamber. (Photos **b** and **c** courtesy of M. Chu)

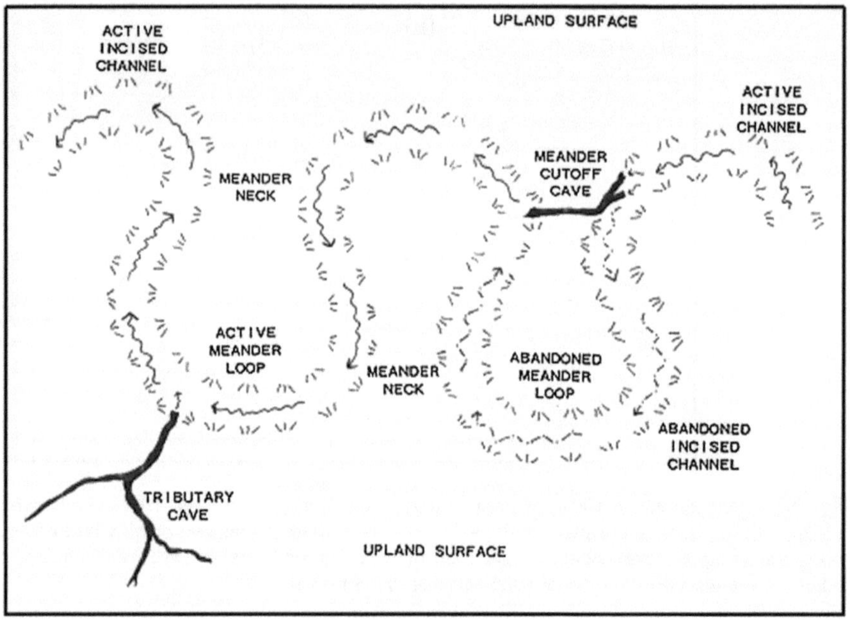

Fig. 9.9 Diagrammatic presentation of stream incision into soluble rock, with subsequent meander cutoff cave development. Note that the meander cutoff cave that has already formed has prevented a second meander cutoff cave from forming at the next downstream meander neck. If a meander cutoff cave forms at the third meander neck (the most downstream one shown), it will influence the tributary cave (From Mylroie and Mylroie 1990)

inventory project of the Adirondack Grotto of the National Speleological Society (Mylroie 1979), and a report with sketch maps by Bob Carroll in 1969 (Carroll 1969).

The key problem, alluded to earlier regarding a post-glacial origin, is that the main cave at the site, Natural Bridge Cave, is quite large and voluminous (Fig. 9.12). Passage widths up to 20 m (65 ft) with heights of 2–5 m (6–16 feet) extend for a linear distance of 175 m (570 ft) to a sump, about two-thirds of the way across the meander neck (Figs. 9.10 and 9.11a). Given that the local drainage appears deranged, and that the caves are quite shallow, these observations support a post-glacial origin. The large passage size suggests Natural Bridge Cave is pre-glacial. The answer may lie in the very high discharges that occur during flood events. Cooper and Mylroie (2014) were able to demonstrate that the nearby maze caves along the banks of the Black River (Chap. 7) formed at an extremely fast rate due to the large

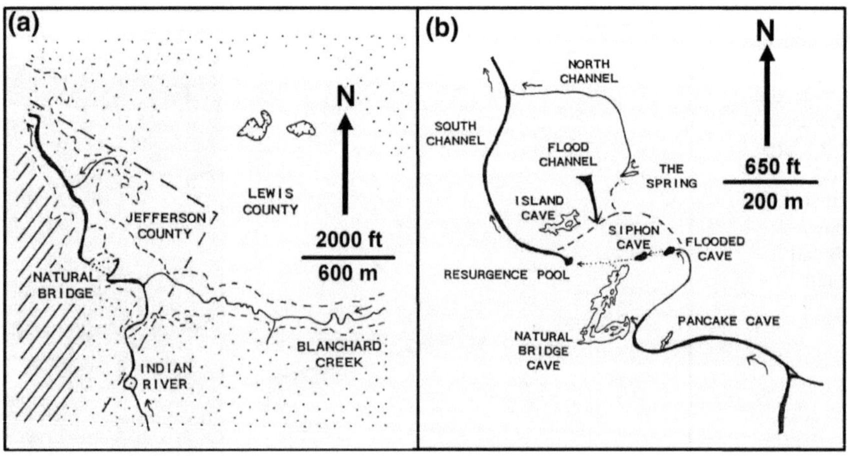

Fig. 9.10 Natural Bridge, Jefferson County, New York. **a** Area map showing glacial sediments in diagonal lines, non-carbonate metamorphic rocks as stippling, and marble areas as *white*. Surface stream flow shown by *arrows*. **b** Map of the Indian River meander cutoffs and associated caves (From Mylroie and Mylroie 1990)

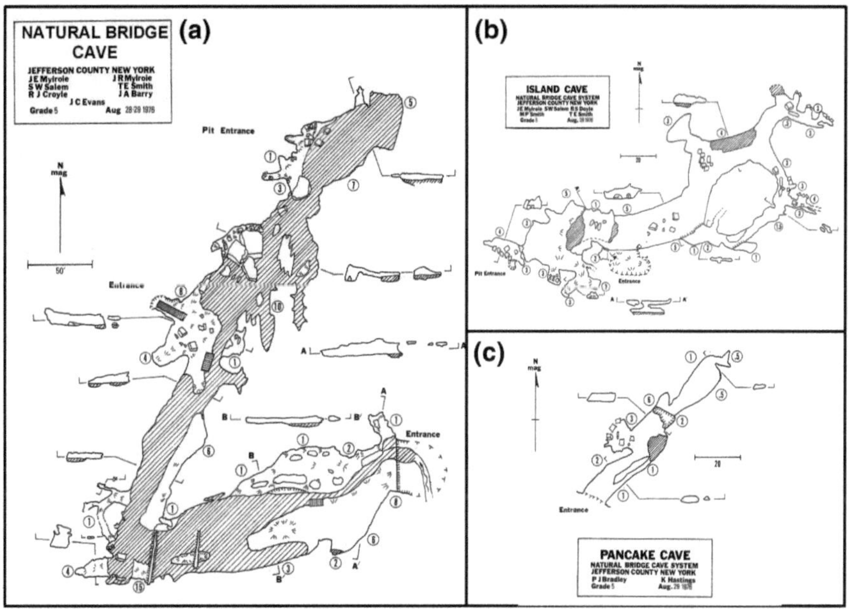

Fig. 9.11 Natural Bridge caves, Jefferson County, New York. All map dimensions in feet (from Mylroie 1979). Refer to Fig. 9.10 for cave positions. **a** Map of Natural Bridge Cave, an active meander cutoff cave; scale bar is 50 ft (15 m). **b** Map of Island Cave, an abandoned meander cut off cave; scale bar is 20 ft (6 m). **c** Pancake Cave, a tributary cave not associated with meander cutoff development; scale bar is 20 ft (6 m)

discharges available. Just as those maze caves formed, and are now abandoned, in the post-glacial time window, the hydrologic history of Natural Bridge Cave is similarly complex. At low flow, the Indian River flows part way around the meander to sink in Flooded Cave as well as Natural Bridge Cave, and transits via Siphon Cave past the downstream sump of Natural Bridge Cave to the resurgence (Fig. 9.10). At higher flows, the Natural Bridge Cave is utilized, and at extremely large flows, a flood channel occupying the original meander is utilized. Island Cave (Figs. 9.10, 9.11b and 9.12b) is dry and abandoned at normal flows, but its position along the original meander suggests it once carried all or part of the Indian River. Its presence suggests that a complex series of stream captures and cave abandonment has occurred. The setting contains several tributary caves, including Pancake Cave (Figs. 9.10 and 9.11c). To compress these events into a 13,000-year time window since de-glaciation is problematic. Engel (2005) has also called for cave development within the post-glacial time window at Natural Stone Bridge and Caves on the east side of the Adirondacks, where a significant stream sinks underground during average flows into large cave passages.

It is possible that Natural Bridge Cave initiation began during ice cover conditions, as part of the basal hydrology of the ice sheet, especially during the retreat phase. Or the cave could be pre-glacial and its presence helped establish the post-glacial surface drainage pattern. But these ideas are pure speculation, as no data, such as pre-glacial or sub-ice/sub-lake sediment deposits exist to support either case. The congruence of the caves today, with the current deranged drainage, and their shallow position in residual hills of marble, argue for *both* a post-glacial origin, and the power of high discharge regimes to create large cave passages quickly.

9.6 Stream Capture in the Grenville Marble

As described above, Natural Stone Bridge and Caves, in Pottersville, Warren County, New York, on the east side of the Adirondacks, is a significant cave compared to the average marble cave in the Adirondacks. Being a show or commercial cave, it has excellent access to a series of short cave passage segments, some of which have large cross sections (Figs. 9.13 and 9.14). The main sink-point entrance is 50 m (165 ft) wide and 10 m (33 ft) high (Fig. 9.14b). The cave has formed in an isolated pod of Grenville Marble, and captures the flow of Trout Brook, which has a catchment of 230 km^2 (90 mi^2), and which bypasses the cave (Fig. 9.14d) during high flow events via an overflow surface channel (Engel 2005). While the straight-line underground traverse is short, less than 200 m (650 ft), a series of passages have developed that capture Trout Brook first on the north bank, then

Fig. 9.12 Natural Bridge Caverns, Jefferson County, New York, as seen in 1976. **a** Ticket Office; Indian River and the Natural Bridge Cave are behind and below the office. **b** Island Cave (Fig. 9.11b), looking west to a collapse entrance. **c** Upstream tourist entrance to Natural Bridge Cave (Fig. 9.11a); note low-flow stream entering the cave under the footbridge. **d** Main Passage of Natural Bridge Cave, with stream trending to the right. Tourists rode small skiffs on the ponded portion of this stream to an exit a few tens of meters away (see Fig. 9.11a). Passage size is very large for a supposed post-glacial cave

again on the south bank (Fig. 9.13), with all water resurging along the downstream north bank of the overflow channel, at a series of caves and springs to the east (Fig. 9.14d).

As noted above, the large size of the cave passages could indicate a pre-glacial origin, but Engel (2005) argues the cave is post-glacial, based on its position in the deranged drainage of the area, and on dissolution rates based on the large discharge available given Trout Brook's catchment size. Engel (2005) also comments on water chemistry data supporting a post-glacial origin, Trout Brook being a classic example of allogenic catchment (sensu Palmer 2007). This interpretation fits with the Cooper and Mylroie (2014) approach to New York maze caves, and would fit with the Natural Bridge Cave area origin on the west side of the Adirondacks, presented earlier.

Crane Mountain Cave, near the foot of Crane Mountain in Warren County, New York, is another stream capture cave in an isolated pod of Grenville Marble (Fig. 9.7). The catchment area and subsequent stream discharge is less, compared to nearby Natural Stone Bridge and Caves, and the cave passage cross section is consequently less as well. The cave is surprisingly complex over its short traversable distance, with upper levels and a large chamber. In a similar fashion, Burroughs Cave (Fig. 9.8) also captures a small stream, but the cave interior is large by New York marble cave standards, and contains a very large room with significant breakdown (Fig. 9.8b). Unlike Natural Stone Bridge and Caves, discharge values for Crane Mountain Cave and Burroughs Cave may not have been large enough to promote rapid cave development in the 13,000-year post-glacial time window. The difficulty in the present-day viewpoint is that the discharges of today may not be those of the past. Both Faulkner (2009) and Quick (2010) argue for a quick initial cave genesis to traversable size as a result of glacial and post-glacial lake flooding and discharge events. Cooper and Mylroie (2014) argue for high discharge conditions as the post-glacial deranged drainage undergoes its first initial adjustments to de-glaciation. Data are lacking to determine how discharge may have changed over the last 13,000 years at Crane Mountain Cave and Burroughs Cave; even historical discharge data are incomplete.

The large breakdown chamber in Burroughs Cave is one of the largest marble cave chambers in the Adirondacks. Explaining its existence as part of post-glacial cave genesis is difficult, as the current drainage

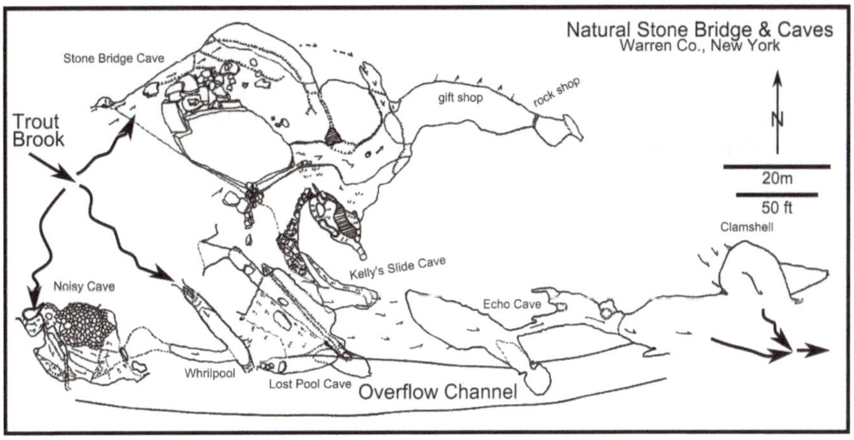

Fig. 9.13 Natural Stone Bridge and Caves, Warren County, New York, showing how Trout Brook distributes itself to several underground flow paths. (From Quick 2010)

Fig. 9.14 Natural Stone Bridge and Caves, Warren County, New York. **a** Upstream Trout Brook at low flow as it reaches the marble outcrop (foreground). **b** Main sink point of Trout Brook into a 30 m wide entrance. **c** Upper entrance to an overflow passage, just to the right and downstream of **b**. **d** Overflow channel with two resurgences, one in the foreground and one in the background, both on the *left* (north) wall of the channel as viewed

does not seem to have a discharge capable of removal of such a large rock volume by dissolution. An alternate explanation may be possible. It has been shown by workers in other localities that when soluble rocks rest on insoluble, resistant rocks, such as a granite at the Kelly Hill Caves on Kangaroo Island, Australia (Jennings 1971) or a basalt on Bermuda (Mylroie et al. 1995), small vadose streams will migrate across that hard contact, undermine the cave walls, and create a broad bedrock span susceptible to collapse (when shales are the underlying insoluble rock, the stream tends to incise into those weak rocks and become fixed in position, as occurs at the "contact" caves of West Virginia; Palmer 2007). If the cave stream in Burroughs Cave (Fig. 9.8c) had a similar lateral migration over underlying insoluble metamorphic rocks, then the collapse could have been initiated. The stream could then have carried off collapse material both by mechanical transport and by dissolution, creating accommodation space for further collapse and thereby creating a large chamber. This explanation is highly speculative, but allows Burroughs Cave to be shoehorned into a post-glacial model of speleogenesis; or the cave could be a "combination" cave, as described by Faulker (2009; see also Chap. 8), a pre-glacial cave utilized by post-glacial hydrology.

9.7 Summary

The studied caves of the Adirondack Mountains seem to mostly fit within post-glacial conceptual models. The maze caves of the Adirondacks are not located either at high elevations that protect them, or in marbles of low lateral extent such as the stripe karst of Norway, making their survival through the previous glaciation unlikely (Cooper and Mylroie 2014). Other caves fit within the TDMO model (Top-Down, Middle-Outwards) given by Faulkner (2008), where caves form rapidly in the post-glacial landscape through high hydraulic gradients combined with mechanically enlarged fractures, followed by conduit-full conditions and vadose enlargement. Both of these styles of cave development leave caves vulnerable to truncation or complete removal upon subsequent glaciation (Faulkner 2008; Cooper and Mylroie 2014). The meander cutoff caves in the Grenville Marble near Natural Bridge, NY are also likely post-glacial with their congruence with post-glacial drainage of the Indian River (Mylroie and Mylroie 1990). These caves therefore work within the early thoughts on glaciated karst; caves of the Adirondack Mountains and within the Grenville Marble are post-glacial in origin, with few surviving pre-glacial cave systems.

The seemingly entirely post-glacial origins for caves within the Adirondack Mountains and Grenville Marble may however be an artifact of this poorly studied region. The knowledge base built in the Helderberg Plateau revealing a mixture of pre-glacial caves with post-glacial modification, and entirely post-glacial caves was a result of budding new cave scientists in the late 20th century in an easily accessible caving region with few complications from structural deformation or metamorphism. The lack of such studies in the complex setting of the Adirondack Mountains may be the source of bias towards post-glacial cave origin explanations in this region.

Another result of studies within the Adirondack Mountains and caves within the Grenville Marble is the seeming unimportance of lithologic controls. The maze caves in the Adirondack Mountains have likely formed in the time since deglaciation, with no potential preferential preservation as seen in the stripe karst maze caves of Norway. The meander cutoff caves near Natural Bridge, New York also do not show developmental differences from those of Kentucky in flat lying, non-deformed limestones. This result is an interesting one as it shows that recharge characteristics and dominant porosity types play more of a role in speleogenesis for this region rather than lithologic controls that affect marble cave development in other glaciated environments such as Norway.

References

Carroll RJ (1969) Some Grenville Marble caves in the northwestern Adirondacks. Northeas Caver 1:114–117

Cooper MP, Mylroie JE (2014) Post-glacial speleogenesis: verification of a hypothetical model, and the origins of maze caves in glaciated terrains. Cave Karst Sci 41(2):84–95

Engel TD (1989) Big loop cave. Northeas Caver 20(1):31–35

Engel TD (2005) Geology of Natural Stone Bridge and Caves, Pottersville, New York. National Speleological Society Convention Program Guide, pp 81–82

Evans J, Quick P, Sloane B (1979) An introduction to caves of the Northeast. National Speleological Society, Huntsville

Faulkner T (2008) The top-down, middle-outwards model of cave development in central Scandinavian marbles. Cave Karst Sci 34(1):3–16

Faulkner T (2009) Speleogenesis in New England marble caves. In: Proceedings of the Fifteenth International Speleological Congress, Kerrville, vol 2, pp 855–862

Isachsen YW, Landing E, Lauber JM et al (2000) Geology of New York: a simplified account, 2nd edn. New York State Museum, Albany

Jennings JN (1971) Karst. MIT Press, Cambridge

Lauritzen S-E, Skoglund RØ (2013) Glacier ice-contact speleogenesis in marble stripe karst. Treatise Geomorphol 6:363–396 (Schroder JF (ed))

Mylroie JE (1979) Geology of the Natural Bridge Cave System: Special Publication Number 1. Adirondack Grotto, Sauquoit

Mylroie JE, Mylroie JR (1990) The consequences of meander incision into soluble rocks. Natl Speleol Soc Bull 52 (1):33–44

Mylroie JE, Carew JL, Vacher HL (1995) Karst development in the Bahamas and Bermuda. In: Curran HA, White B (eds) Geological society of America special paper 300, Terrestrial and shallow marine geology of the Bahamas and Bermuda, pp 251–267

Nardacci M (ed) (1991) Guide to the caves and karst of the Northeast. National Speleological Society, Huntsville

Palmer AN (2007) Cave geology. Cave Books, Dayton

Quick PG (2010) Vermont caves: a geological and historical guide. National Speleological Society, Huntsville

Ridge JC (2004) The Quaternary glaciation of Western New England with correlations to surrounding areas. In: Ehlers J, Gibbard PL (eds) Quaternary glaciations—extent and chronology—Part II: North America. Developments in quaternary sciences, vol 2. Elsevier, Amsterdam, pp 169–199